Models of
Network Reliability

Analysis, Combinatorics,
and Monte Carlo

Models of
Network Reliability

Analysis, Combinatorics, and Monte Carlo

Ilya B. Gertsbakh • Yoseph Shpungin

CRC Press
Taylor & Francis Group
Boca Raton London New York

CRC Press is an imprint of the
Taylor & Francis Group, an **informa** business

CRC Press
Taylor & Francis Group
6000 Broken Sound Parkway NW, Suite 300
Boca Raton, FL 33487-2742

First issued in paperback 2019

ISBN-13: 978-1-4398-1741-4 (hbk)
ISBN-13: 978-0-367-38456-4 (pbk)

Library of Congress Cataloging-in-Publication Data

Gertsbakh, I. B. (Il,ia Borukhovich)
   Models of network reliability : analysis, combinatorics, and Monte Carlo / Ilya B. Gertsbakh, Yoseph Shpungin.
     p. cm.
   Includes bibliographical references and index.
   ISBN 978-1-4398-1741-4 (hardcover : alk. paper)
   1. Network performance (Telecommunication) 2. Reliability (Engineering)--Statistical methods. 3. Computer networks--Reliability. I. Shpungin, Yoseph. II. Title.

TK5105.5956.G487 2010
621.382'1--dc22
                                         2009040273

Visit the Taylor & Francis Web site at
http://www.taylorandfrancis.com

and the CRC Press Web site at
http://www.crcpress.com

To my wife Ada and my daughter Janna
I.G.

To my wife Elena and my son Hanan
Y.S.

Contents

PREFACE

This is a one-semester graduate course in Network Reliability Analysis, Combinatorics, and Monte Carlo, for Software Engineering, Industrial Engineering, and Computer Science students. We assume that the students have command of Calculus, of a programming language, and of Probability Theory with basics of Statistics.

In Reliability Theory, like in any theory, we think and operate in terms of *models*. A rather popular model for studying computer/communication network reliability is a graph with nodes and/or edges subject to failures. The main object of the study is the so-called K-terminal network reliability, i.e. the probability that a special set of K nodes called *terminals* remain connected with each other.

For "standard" reliability analysis even a small network is a system with a very large number of states. For example, a complete graph with 10 nodes and unreliable edges has 2^{45} different binary states, a huge number, and its "straightforward" reliability calculation is impossible.

Met with this and similar situations, researchers' efforts were directed toward using Monte Carlo (MC) simulation. MC method is a specially designed computerized statistical experiment by means of which it becomes possible to obtain an estimate of the reliability parameters of interest. MC methods usually do not work well unless they exploit the specific properties of the object under study.

For networks and their reliability parameters, most efficient MC algorithms use such notions as, e.g. minimal/maximal network spanning trees, minimal paths and minimal cuts, so-called network destruction/construction spectra. The power of these algorithms lies in the fact that these network combinatorial parameters characterize *topological* properties of the network and do not depend on reliability characteristics of the network's unreliable components.

Chapter 1 is an elementary introduction to MC methodology. It is written for readers who have no previous experience in MC methods. It has also a short reminder of basic facts from Statistics needed for understanding the future material.

Chapter 2 presents a short review of network types, network topology, and basic reliability notions of networks and describes standard techniques for their reliability evaluation. Special attention is paid to constructing network spanning trees, one of the main combinatorial instruments for further reliability analysis.

One of the most powerful methods of network reliability estimation is a MC scheme called "Lomonosov's turnip", invented in 1974 [36] and implemented in 1991 [11]. It uses an artificial edge evolution process in which each network's edge is supplied with a random time of its "birth". This time must be distributed according to the exponential distribution, due to its unique properties. Chapter 3 studies this distribution and presents simple random processes related to it.

Chapter 4 is a further excursion into network reliability theory and its central topic is so-called Burtin-Pittel (BP) approximation to network failure probability [6]. BP method is based on network minimal cut sets and produces good approximation to reliability parameters for highly reliable networks.

Chapter 5 analyzes networks whose elements have random lifetimes. It turns out that some "pre-failure" states (so-called "border" states) are of special interest since they are related to network reliability gradient function. This function will be used further in connection to network component importance and its synthesis.

When network nodes or edges fail in time, of greatest interest becomes the ordinal number of the "critical" component whose failure signifies the whole network failure. The lifetime distribution of the critical component is described in terms of order statistics. In Chapter 6 we remind the basic definitions of order statistics and describe the distribution of the "critical" number called *spectrum*, which becomes a very important network reliability combinatorial characteristic.

Chapter 7 is devoted to a rather special topic and can be omitted in the first reading of the book. It presents a MC algorithm for estimating convolutions, an issue which becomes acute in implementing Lomonosov's turnip, since the analytic methods can produce low accuracy and accumulation of errors.

Chapters 1-6 can serve as an independent course in reliability of network-type systems with some elements of their MC simulation methodology.

The principal MC algorithms are described in Chapters 8-12. Chapter 8 presents a very efficient MC algorithm for estimating network lifetime in the course of its destruction (when edges or nodes fail in time). This algorithm is based on the properties of the maximal spanning tree of the network.

Chapter 9 contains a detailed description of so-called "Lomonosov's turnip" algorithm and its applications. This algorithm is our basic tool for calculating the probability of network K-terminal connectivity, for both unreliable nodes or unreliable edges. In addition, this algorithm allows to estimate the stationary mean UP and $DOWN$ periods for networks with renewable components. A slight modification of Lomonosov's algorithm makes it possible to compute the network reliability gradient vector.

One of the first steps in reliability theory was introducing the notion of component importance, so-called Birnbaum Importance Measure (BIM) [3]. To improve system reliability, the "most important" component is the first candidate for network's reinforcement. In reliability practice, the implementation of importance measures even for small networks was minimal since it demanded the knowledge of the analytic formula for system reliability. Chapter 10 overcomes this obstacle by introducing an efficient MC algorithm for estimating BIM's for relatively large networks.

Chapter 11 is devoted to a relatively new topic in reliability theory - the optimal network synthesis. It deals with the "best possible" allocation of reliability resources into network components in order to design a network with optimal value of its reliability.

Chapter 12 is devoted to a new and rather complicated issue in network reliability - the study of the network exit time. It describes how to use the Lomonosov's algorithm to construct accurate bounds on the distribution function of the exit time from network UP state into its $DOWN$ state.

Chapter 13 presents reliability simulation results for a collection of computer networks. It is concluded by a realistic example of an integrated computer-communication network with unreliable nodes. On this example we demonstrate how to apply our analytic, combinatorial, and MC tools developed in previous chapters to obtain a comprehensive system reliability analysis, locate the "soft" spots of the system and suggest their optimal reinforcement strategy.

Each chapter is concluded by several problems and exercises which are intended to help the students to get better command of the material. A few topics, like the use of $O(\cdot)$ and $o(\cdot)$ symbols, are explained in Appendices. Of special importance is Appendix C which contains a unique simulated collection of destruction spectra for reliability analysis of some typical popular network structures, such as complete graphs, hypercubes, and butterfly networks.

We tried to reduce formal proofs to a minimum. When the proof was necessary we tried first to present its idea using examples and intuitive reasoning. During the first reading we advise the readers to concentrate more on numerical examples and definitions and less on theorems and their proofs. In order to make easier the exposition, we often repeat the previously given facts and/or definitions to avoid from the reader the necessity to search the preceding material.

As it happens in almost every field, the true "feeling" of the material and its good understanding comes after solving problems and exercises and implementing the algorithms in computer programs. We strongly recommend the students to develop their own computer programs for calculating network reliability, even if the programs use the simplest and not the most sophisticated algorithms. A good start might be writing a computer code for crude Monte Carlo to estimate network reliability, using the algorithms described in detail in Chapter 2. We believe that all students taking this course will have necessary elementary command of computer programming which would enable them to take an active part in designing various programs for network reliability evaluation. Heavy issues like the "turnip" algorithm and its modifications might be an excellent project theme for this course.

It is a pleasure to acknowledge our long-standing indebtedness to Michael Lomonosov, who introduced us to network combinatorics, with his characteristic imagination and original insight.

I. Gertsbakh, Y. Shpungin

elyager@bezeqint.net   yosefs@sce.ac.il

Beersheva - Tel-Aviv

Notation and Abbreviations

i.i.d. - independent identically distributed.

i.r.v. - independent random variables.

r.v. - random variable.

c.d.f., CDF - cumulative distribution function.

d.f. - density function.

τ, X, Y, Z - random variables.

$X \sim Exp(\lambda)$ - r.v. X is exponentially distributed with parameter λ.

$X \sim U(0, 1)$ - r.v. X is uniformly distributed on [0,1].

$E[X]$ - mathematical expectation (mean value) of r.v. X.

$Var[X]$ - variance of r.v. X.

σ - square root of variance, $\sigma = \sqrt{Var[X]}$, also termed standard deviation.

σ_X - standard deviation of r.v. X.

$X \sim B(n, p)$ - r.v. X has a binomial distribution: X is the total number of successes in n independent experiments with probability p of success in a single experiment.

$\tau \sim Exp(\lambda)$ - r.v. τ has exponential distribution with parameter λ.

$X \sim Gamma(n, \lambda)$ - r.v. X has gamma distribution with parameters n, λ.

$X \sim Poisson(\Lambda)$ - r.v. X has Poisson distribution with parameter Λ.

$X \sim F(t)$ - r.v. X has CDF $F(t)$, i.e. $F(t) = P(X \leq t)$.

$r.e.[X]$ - relative error of r.v. X. Defined as $\sigma_X / E[X]$, for nonnegative r.v.'s only.

$X \sim \diamond(\mu, \sigma)$ - r.v. X has mean value μ and st.dev σ.

ST - spanning tree.

MST - minimal spanning tree.

MaxST - maximal spanning tree.

$Max^{(min)}ST(T)$ - minimal subtree of the maximal spanning tree of terminal set T.

BIM - Birnbaum importance measure.

CMC - Crude Monte Carlo.

MC - Monte Carlo.

S_p - destruction spectrum, or D-spectrum.

S_c - construction spectrum or C-spectrum.

Chapter 1

What is Monte Carlo Method?

A pint of example is worth a gallon of advice.

Quips and Quotes

From *Cambridge Dictionary of Statistics* [13]:

Monte Carlo Methods:

Methods for finding solutions to mathematical and statistical problems by <u>simulation</u>. Used when the analytic solution of the problem is either intractable or time consuming.

Simulation:

The artificial generation of random processes (usually by means of *pseudorandom* numbers and/or computers) to imitate the behavior of particular statistical models. See also *Monte Carlo Methods*.

Let us consider several examples.

1.1 Area Estimation

Suppose we have a figure A whose area S is difficult to calculate analytically. The figure is placed into a unit square, see Fig. 1.1.

First, a basic fact. Let $X \sim U(0,1)$ and $Y \sim U(0,1)$ be i.r.v.'s. Then the probability that a random point $Z = (X, Y)$ "thrown" on the unit square

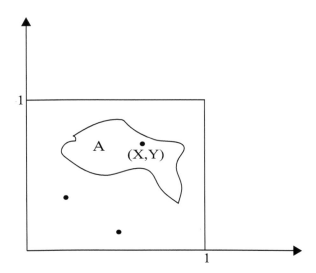

Figure 1.1: Figure A in a unit square

falls into A is equal to the area S of A. Formally,

$$P(Z \in A) = S. \tag{1.1.1}$$

Define a binary r.v. V equal to 1 if $Z \in A$, and 0 otherwise. Obviously, $E[V] = S, \ Var[V] = S(1 - S)$.

Algorithm 1.1 - AREACRUDE
1. **Put** $AREA$:=0.
2. **Generate** X, Y, both independent and uniformly distributed on [0,1].
3. **If** $Z = (X, Y) \in A$, **set** $AREA := AREA+$ 1.
4. **GOTO** 2.
   Repeat 2-4 N times
   Estimate area S by

$$\hat{S} = AREA/N. \tag{1.1.2}$$

In fact, the estimate of S is the fraction of random points which fell into A. Now it is easy to check that $E[\hat{S}] = S$, i.e. $\hat{S}$ is an unbiased estimator of S. Moreover, it is easy to check (do it!) that

$$Var[\hat{S}] = S(1 - S)/N. \tag{1.1.3}$$

As $N \to \infty$, $Var[\hat{S}] \to 0$. Thus $\hat{S}$ converges to the true value of the area S.

Comment. If you have difficulty with checking the last formula, recall that for i.r.v.'s $V_1, V_2, ..., V_m$,

$$Var[a_1 V_1 + ... + a_m V_m] = \sum_{i=1}^{m} a_i^2 Var[V_i].\#$$

The above example is quite simple and the algorithm does not seem to create any problems. However, it turns out that the *relative error* (*r.e.*) of the estimate $\hat{S}$ becomes very large if the area S becomes very small.

The *r.e.* is, by the definition, the ratio of the standard deviation of X and its mean, i.e.

$$r.e.[X] = \frac{\sqrt{Var[X]}}{E[X]} = \sigma_X / E[X]. \tag{1.1.4}$$

(The relative error is defined only for nonnegative random variables.) For a binary random variable $X \sim B(1, p)$,

$$r.e.[X] = \frac{\sqrt{p(1-p)}}{p}. \tag{1.1.5}$$

Applying (1.1.4) to the r.v. $\hat{S}$ we obtain

$$r.e.[\hat{S}] = \frac{\sqrt{1-S}}{\sqrt{S \cdot N}}. \tag{1.1.6}$$

Obviously, for fixed N, $r.e.[\hat{S}]$ goes to infinity as S tends to zero. In practical terms it means that the Monte Carlo estimator of S for small S is highly inaccurate. We will see later how certain Monte Carlo schemes can be modified to preserve bounded value of the *r.e.*

Remark

Suppose that p is close to one. Then we will be interested in accurate estimation of $q = 1 - p$. The appropriate expression for the relative error will be $\sqrt{p(1-p)}/q$, and this expression tends to infinity as $q \to 0$.

1.2 Optimal Location of Components

Eight components should be positioned on a planar heat conducting board, see Fig. 1.2 below.

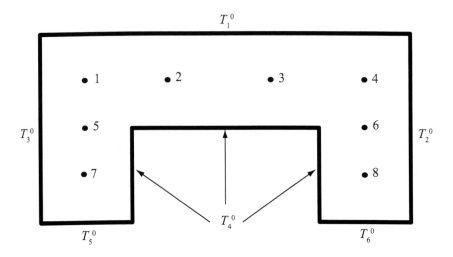

Figure 1.2: The board and the components. T_i^o are the border temperatures

The periphery of the board is kept at fixed temperature by using special cooling devices. Component i dissipates power w_i, and all components together create, in stationary regime, a thermal field on the board with temperature T_j at the j-th position. If the location of components changes, the temperature field on the board also changes. Each component has its own *failure rate* ψ depending on the temperature at the point where the component is positioned. The total system failure rate is defined as

$$\Lambda(T_{i(1)}, ..., T_{i(8)}) = \sum_{j=1}^{8} \psi(i(j), T_{i(j)}), \qquad (1.2.1)$$

where j is the position number, i is the component number, and $T_{i(j)}$ is the temperature at the j-th position where the i-th component is located.

Our purpose is to find the optimal location of all components which would minimize the value of Λ. There are 8!=40320 different component locations. Although this number is not very large, the following algorithm which uses random search may work quite well even for considerably larger number of components.

Denote by $\pi = (i_1, i_2, ..., i_8)$ the permutation describing the location of components on the board: position $k, k = 1, ..., 8$, is occupied by component i_k.

Algorithm 1.2 - RANDOMSEARCH

1. **Choose** a permutation π.
2. **Compute** the stationary temperature field. **Denote** it by $\mathbf{T}(\pi)$.
3. **Compute** the total system failure rate $\Lambda(\mathbf{T}(\pi)) = F_0$.
4. **Pick** randomly two components and **exchange** them, i.e. component on position i is put on position j and vice versa. Denote by π^* the new permutation after this exchange.
5. **Calculate** the new temperature field and the corresponding $\Lambda(\mathbf{T}(\pi^*)) = F_1$.
6. **If** $F_1 < F_0 - \epsilon$, **put** $F_0 := F_1, \pi := \pi^*$ and GOTO **4**.
   If $F_1 \geq F_0 - \epsilon$, try another random exchange.
   If K random exchanges do not result in the decrease of Λ by more than ϵ, then **STOP**.

In simple words, we are trying to improve the existing permutation π by pairwise random exchanges of components, and do it at most K times. If we don't achieve decrease in the failure rate by more than some small ϵ, we stop and declare the present permutation as "optimal". In practice, $K = 10 - 30$.

In numerical experiments we typically observe the following behavior of Λ, see Fig. 1.3:

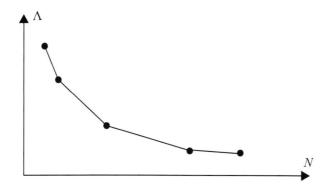

Figure 1.3: N is the total number of permutations. Λ initially decreases rapidly and later stabilizes as N grows

The above random search has several modifications. One of them avoids stopping in a local minimum which is not the global minimum. If the algorithm stops at a permutation which cannot be improved by K pairwise exchanges, the algorithm accepts a permutation which is locally *worse* than

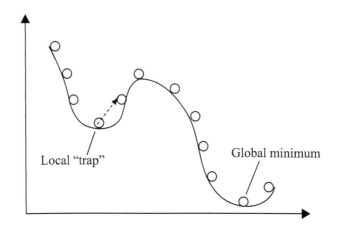

Figure 1.4: The arrow shows a step in "wrong" direction

the existing one. This might avoid stopping the algorithm in a local minimum which is not a global one. This situation is illustrated by Figure 1.4.

1.3   Reliability of a Binary System

Suppose we have a bridge type structure (see Fig. 1.5). Edges 1,2,3,4,5 are subject to failures. The bridge is "*UP*" if there is a connection between S and T, and "*DOWN*", otherwise. Our goal is to estimate R, the probability that the bridge is "*UP*".

About the component failure probabilities we know the following: in a "good" day the failure probabilities for all components are equal 0.01, and on the "bad" day, the failure probabilities for the components are 0.02. The proportion of the bad days is 20%. Our purpose is to estimate the average system reliability in a randomly chosen day.

Algorithm 1.3 - AverageReliability
1. **Put** $R:=0$.
2. **Simulate** random variable DAY which is equal G with probability 0.8 and B with probability 0.2.
3. **If** DAY=G, **Simulate** r.v.'s $X_i \sim B(1, q = 0.01)$, for $i = 1, ..., 5$.
   If DAY=B, **Simulate** r.v.'s $X_i \sim B(1, q = 0.02)$, for $i = 1, ..., 5$.
4. **Erase** each edge i if $X_i = 0$.

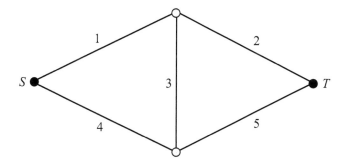

Figure 1.5: Bridge structure

5. Set $I:=1$ if the bridge is UP and $I:=0$, otherwise.
6. Put $R := R + I$.
7. Repeat steps 2-6 N times.
8. Estimate the system reliability R as $\hat{R} = R/N$.

1.4 Statistics: a Short Reminder

1.4.1 Unbiased estimators

Let us have a closer look at the result of the Monte Carlo simulation experiment, for example, the area estimation by means of AREACRUDE algorithm in the form of ratio $\hat{S} = AREA/N$, see (1.1.2). In this formula, $AREA$ is the result of a random experiment. In a particular experiment, we observe one particular replica of this random variable. Let us have a closer look at this random variable. It is seen from the above algorithm that

$$AREA = V_1 + V_2 + ... + V_N, \qquad (1.4.1)$$

where $V_i, i = 1, ..., N$, are independent identically distributed (i.i.d) 0-1 random variables. Such r.v.'s are called *binary*. $P(V_i = 1) = p, P(V_i = 0) = 1 - p$, where p is the probability that the random point (X, Y) falls into the area A shown on Fig. 1.1. By the very definition of (X, Y) this happens with probability equal to S which equals the area A. In other words, $p = $ Area of A $= S$.

Since the result of our MC simulation experiment is *random* we are interested in the statistical properties of this experiment. In particular, we want to know how close is the simulation result $\hat{S}$ to the true value of the area A. This leads us to the investigation of the mean value and the variance of the r.v. $\hat{S}$. We will do it in a slightly more general situation in which the quantity of interest $\hat{\theta}$ is represented as

$$\hat{\theta} = \frac{X_1 + X_2 + \dots + X_N}{N}, \tag{1.4.2}$$

where $X_i, i = 1, 2, \dots, N$ are i.i.d. random variables, not necessary of binary type. All we know about these r.v.'s is that they have finite mean value $E[X_i] = \mu$ and variance $Var[X_i] = \sigma^2$. We remind the reader that, physically speaking, the mean value is the *center* of the probability mass of the random variable, and the variance characterizes the *spread* of the probability mass around this center. More formally, it is useful to remember that

$$Var[X] = E[(X - \mu)^2] = E[X^2] - E^2[X]. \tag{1.4.3}$$

The first fundamental fact is the following property of the mean value of the linear combination of random variables. It says that the mean of the linear combination of r.v.'s is a linear combination of the means of these r.v.'s. Important is that this remains true no matter if the random variables are independent or not. Formally, for any collection of numbers $a_1, a_2, \dots, a_N$,

$$E[\sum_{i=1}^{N} a_i \cdot X_i] = \sum_{i=1}^{N} a_i \cdot E[X_i]. \tag{1.4.4}$$

Now let us apply (1.4.4) to the above expression of $\hat{\theta}$ (1.4.2). Note that now all X_i are i.i.d., $E[X_i] = \mu$ and all $a_i = 1/N$. Therefore,

$$E[\hat{\theta}] = \sum_{i=1}^{N} \mu/N = \mu. \tag{1.4.5}$$

Thus, the mean value of $\hat{\theta}$ coincides with the mean value of r.v. X_i. In statistics we say that $\hat{\theta}$ is an *unbiased estimator* of μ.   In simple words, the result of our random experiment, *on the average*, is equal to the mean value of the random variable X_i. Let us check what this property gives us in the particular case of area estimation.

$$E[\hat{S}] = E[AREA/N] = E[(V_1 + V_2 + \dots + V_N)/N] = \tag{1.4.6}$$

$$\sum_{i=1}^{N} E[V_i]/N = p = \text{Area of A} = S.$$

So, $\hat{S}$ is an unbiased estimator of the area S.

1.4.2 Variance behavior of an estimator as sample size increases

Unbiasedness itself is not enough to provide an estimator of a good quality. We must check what happens with its variance, and the highly desirable property would be having a variance which is getting smaller as the number of experiments N in the simulation experiment gets larger. To verify how the variance of $\hat{\theta}$ behaves let us remind (without proof) the following property of the variance of the linear combination of random variables. Let

$$Y = a_1 X_1 + a_2 X_2 + ... + a_n X_N. \tag{1.4.7}$$

Then

$$Var[Y] = \sum_{i=1}^{N} a_i^2 \cdot Var[X_i] + 2 \sum_{i \neq j, i < j} Cov[X_i, X_j] \cdot a_i a_j, \tag{1.4.8}$$

where $Cov[X_i, X_j] = E[X_i \cdot X_j] - E[X_i] \cdot E[X_j]$.

In most cases, r.v.'s X_i in (1.4.7) will be independent and identically distributed. Independence implies that the covariances will be equal zero, which is a great relief. Identical distributions imply that all variances are the same, $Var[X_i] = \sigma^2$. Then (1.4.8) simplifies to

$$Var[Y] = \sigma^2 \cdot \sum_{i=1}^{N} a_i^2. \tag{1.4.9}$$

Suppose now that all $a_i = 1/N$. Then

$$Var[Y] = \frac{\sigma^2}{N}. \tag{1.4.10}$$

The presence of N in the denominator is of crucial importance. If we construct our estimator in a form similar to the form of Y with i.i.d. X_i-s and $a_i = 1/N$, the variance of our estimator will tend to zero as $N \to \infty$.

Let us check what happens with the variance of our area estimator $\hat{S}$.

$$Var[\hat{S}] = \frac{Var[V_1] + ... + Var[V_N]}{N^2} = \frac{Var[V_i]}{N}. \tag{1.4.11}$$

Since V_i is a binomial random variable which takes 1 and 0 values with probability p and $1 - p$, respectively, its variance equals

$$Var[V_i] = p(1 - p). \tag{1.4.12}$$

Finally,

$$Var[\hat{S}] = \frac{p(1-p)}{N}. \tag{1.4.13}$$

Therefore, our MC estimator of the area is not only unbiased, but also has variance tending to zero as the number of experiments increases. In other words, with the increase of N, the value of $\hat{S}$ becomes closer and closer to the true (unknown) value of the area S.

In MC practice, we often use so-called *relative error* (r.e.) defined as the ratio of the square root of the variance of the estimator $\hat{\theta}$ to its mean value:

$$r.e.[\hat{\theta}] = \frac{\sqrt{Var[\hat{\theta}]}}{E[\hat{\theta}]}. \tag{1.4.14}$$

For the case of area estimation we have

$$r.e.[\widehat{AREA}] = \frac{\sqrt{1-p}}{\sqrt{N \cdot p}}. \tag{1.4.15}$$

In practice we will be satisfied with an MC experiment if it guarantees small *r.e.*, say of 1%. For example, if we have some *a priori* knowledge of the value of the area S, we can plan in advance the number of simulation trials to reach the desired accuracy.

Let us consider a numerical example. Suppose the area is expected to be about 0.1. We want to guarantee *r.e.* of 5%. Then, solving the previous equation with respect to N we obtain

$$N = \frac{1-S}{S \cdot (r.e.)^2} = 0.9/(0.1 * 0.05^2) = 3600. \tag{1.4.16}$$

Advanced probability and/or statistics courses present proof of the fact that the sample average of the form $\bar{X} = \sum_{i=1}^{N} X_i/N$ tends to $\mu = E[X_i]$ *with probability 1* as $N \to \infty$. In terms of the MC estimators considered above, this fact means that with probability 1 the unbiased MC estimator $\hat{\theta}$ tends to its mean value μ as $N \to \infty$.

We conclude this subsection with rule of thumb about the minimal number of binomial trials which is needed to estimate the unknown probability p with a given relative error.

A rule of thumb: To estimate a small probability p from N binomial trials, with a relative error not greater than 0.1 (10%), it is necessary to carry out at least

$$N = \frac{100}{p}$$

binomial trials.

This rule follows immediately from (1.4.15). So, to estimate $p \approx 0.01$ with 10% *r.e.* we need at least 10,000 trials. To estimate $p \approx 0.0001$, with *r.e.* $= 0.1$, we need at least 1,000,000 trials.

1.4.3 Variance in a multinomial experiment

It will be instructive to present a formula for the variance of an estimator obtained in a *multinomial* random experiment. In a *binomial* experiment we observe one of two possible outcomes. So, a random point (X, Y) either falls inside area A (first outcome) or outside it (the second outcome).

Imagine a random experiment which has n, $n > 2$ possible outcomes. Let us number these outcomes as $1, 2, ..., n$. Outcome i appears with probability $f_i > 0$, and $f_1 + f_2 + ... + f_n = 1$. The simplest way to create a multinomial experiment is to partition the unit square on Fig. 1.1 into n non overlapping parts, to number these parts from 1 to n and to denote by f_i the size of the area of the i-th part. If a random point (X, Y) hits part k, we say that the experiment had the outcome k.

Now suppose that we repeat the above experiment N times and register that the outcome i has appeared N_i times, so that

$$N = N_1 + N_2 + ... + N_n. \tag{1.4.17}$$

Now associate a *reward* a_i (nonrandom) with the i-th outcome. Then the total reward of N multinomial experiments will be

$$R = \sum_{k=1}^{n} a_k N_k. \tag{1.4.18}$$

Obviously, $E[R] = N \sum_{k=1}^{n} a_k \cdot f_k$. (Prove it!)

Let us derive a formula for $Var[R]$. It is a good exercise and its result will be useful for further exposition. To simplify the derivation, let us do it for particular case of $n = 3$. Suppose that we describe the result of the i-th experiment by a three-dimensional vector

$$W_i = (X_i, Y_i, Z_i), \tag{1.4.19}$$

where $W_i = (1, 0, 0)$ if the first outcome takes place, $W_i = (0, 1, 0)$ if the second outcome takes places, and $W_i = (0, 0, 1)$ if the third outcome takes place. Obviously, $X_i + Y_i + Z_i = 1$ and

$$P(X_i = 1) = f_1, P(Y_i = 1) = f_2, P(Z_i = 1) = f_3, f_1 + f_2 + f_3 = 1.$$

Now represent the reward in N experiments as

$$R = \sum_{i=1}^{N}(a_1 X_i + a_2 Y_i + a_3 Z_i). \tag{1.4.20}$$

Since the outcomes of j-th and k-th experiment for $k \neq j$ are independent, variance of the sum is a sum of variances, and

$$Var[R] = \sum_{i=1}^{N} Var[a_1 X_i + a_2 Y_i + a_3 Z_i]. \tag{1.4.21}$$

Now note that $X_i = X_i^2, Y_i = Y_i^2, Z_i = Z_i^2$, and $E[X_i] = f_1, E[Y_i] = f_2, E[Z_i] = f_3$.

Note also that $X_i \cdot Y_i = 0, \ X_i \cdot Z_i = 0, \ Y_i \cdot Z_i = 0$. Now

$$Var[a_1 X_i + a_2 Y_i + a_3 Z_i] = \tag{1.4.22}$$
$$E[(a_1 X_i + a_2 Y_i + a_3 Z_i)^2] - E^2[a_1 X_i + a_2 Y_i + a_3 Z_i].$$

After simple algebra we arrive at the following expression:

$$Var[R] = N(\sum_{i=1}^{3} a_i^2 f_i(1 - f_i) - 2 \sum_{1 \leq i < j \leq 3} a_i a_j f_i f_j. \tag{1.4.23}$$

It is easy to guess the form of a similar expression for an experiment with arbitrary number of outcomes $n > 3$. Try it as an exercise!

1.4.4   Confidence interval for population mean based on the normal approximation

Suppose we are performing a simulation experiment in order to estimate the population mean μ. For this purpose we draw a random sample of N observations $X_1, X_2, ..., X_N$ and write the estimate as

$$\hat{\mu} = \frac{\sum_{i=1}^{N} X_i}{N}. \tag{1.4.24}$$

Suppose that X_i are i.i.d. random variables with mean μ and variance σ^2. Very often we are not satisfied with point estimate $\hat{\mu}$ and the respective relative error and we want to supplement the result of the experiment by an *interval* within which the true value μ lies with guaranteed probability. Such an interval is called a *confidence interval*.

There are several approaches to constructing confidence intervals. The most popular one is based on the fact that the sample mean $\bar{X} = \sum_{i=1}^{N} X_i / N$

for large N is approximately normally distributed with mean μ and standard deviation $\sigma/\sqrt{N}$. This is equivalent to saying that

$$Z = \frac{\bar{X} - \mu}{\sigma/N} \approx N(0,1). \tag{1.4.25}$$

We will not derive this result which follows from a rather involved theory of normal approximation and simply present the formula for the confidence interval. In practice one usually uses the confidence level of 95%, which means that the confidence interval covers the population mean with probability 0.95. The confidence interval has the form $[LB, UB]$, where

$$LB = \hat{\mu} - 1.96 \cdot \hat{\sigma}/\sqrt{N}; \qquad UB = \hat{\mu} + 1.96 \cdot \hat{\sigma}/\sqrt{N}, \tag{1.4.26}$$

where $\hat{\sigma}$ is the following estimate of the standard deviation.

$$\hat{\sigma} = \sqrt{\frac{\sum_{i=1}^{N}(X_i - \bar{X})^2}{N-1}}. \tag{1.4.27}$$

In binomial experiment, the formula corresponding to (1.4.26) is

$$LB = \hat{\mu} - 1.96\sqrt{\hat{p}(1-\hat{p})/N}; \ UB = \hat{\mu} + 1.96\sqrt{\hat{p}(1-\hat{p})/N}. \tag{1.4.28}$$

Example 1.4.1. Suppose, we did $N = 1000$ experiments in estimating the area, as described in Section 1.1. We obtained the estimate $\hat{p} = 0.345$, which means that 345 points out of 1000 fell into A. Then $\hat{\sigma} = \sqrt{\hat{p}(1-\hat{p})/1000} = 0.0150$. Thus, according to (1.4.28) the 95% confidence interval is $[0.32, 0.38]$.#

1.4.5 Confidence interval for the binomial parameter: Poisson approximation

Suppose we carry out a binomial experiment a large number of times N, and the number of cases N_1 in which we observed the result $X_i = 1$ is very small. In other words, our estimate $\hat{p} = N_1/N$ is close to zero. If $\hat{\lambda} = N \cdot \hat{p}$ lies in the interval [1,30], the Normal approximation (1.4.25) may not be accurate, and it is more adequate to use a confidence interval based on the *Poisson approximation* to the distribution of $N \cdot \hat{p}$, see Morris H. DeGroot [10], pp. 256-257.

Table 1.1 presents 0.95 - confidence intervals $[LB, UP]$ on the Poisson parameter λ based on the observed value $\hat{\lambda}$. It is borrowed from L.N. Bol'shev and N.V. Smirnov, *Tables of Mathematical Statistics* [4], pp. 368-369.

Table 1.1: 0.95 -Confidence Intervals for Poisson Parameter

$\hat{\lambda} = N \cdot \hat{p}$	LB	UB	$\hat{\lambda} = N \cdot \hat{p}$	LB	UB
1	0.0253	5.57	12	6.20	20.96
2	0.242	7.22	13	6.92	22.23
3	0.62	8.77	14	7.65	23.69
4	1.09	10.24	15	8.40	24.74
5	1.62	11.67	16	9.15	25.98
6	2.20	13.08	17	9.90	27.22
7	2.81	14.42	18	10.67	28.45
8	3.45	15.76	19	11.44	29.67
9	4.12	17.08	20	12.22	30.89
10	4.80	18.39	25	16.18	26.90
11	5.49	19.68	30	20.24	42.83

The following example explains how to use the table.

Example 1.4.2. Suppose we generated the random point (X, Y) $N = 50000$ times and it hit $N_1 = 15$ times the area A. Thus $\hat{p} = 0.0003$. Entering Table 1.1 with $\hat{\lambda} = N \cdot \hat{p} = 15$, we obtain $[LB = 8.40, UP = 24.74]$, which gives the lower bound on p as $p_{LB} = 8.40/50000 = 0.000168$, $p_{UB} = 24.74/50000 = 0.0004955$.

It is interesting to compare these bounds with the normal approximation case.

By (1.4.28),

$$LB = 0.0003 - 1.96\sqrt{\hat{p}(1 - \hat{p})/50000} = 0.0003 - 0.000155 = 0.000145,$$

$$UB = 0.0003 + 0.000155 = 0.000455.$$

In this case both approximations produce similar results. #

1.5   Problems and Exercises

1. Finding the Confidence Level.

Suppose that you are dealing with quality control and face the following situation. You are taking small samples of $n = 4$ independent observations $X_1, ..., X_4$ from a population with unknown distribution : $X \sim \Diamond(\mu, \sigma)$. You can estimate the population mean μ by the average as $\hat{X} = \sum_{i=1}^{4} X_i/4$ and sample standard deviation σ as

$$s = \sqrt{\sum_{i=1}^{4} (X_i - \hat{X})^2 / 3} \tag{1.5.1}$$

You are interested in obtaining a 90% -confidence interval (CI) on the unknown mean.

You remember from Statistics course that for the normal case the Student's confidence interval is applicable, and it has the form:

$$[\hat{X} - (s/\sqrt{4}) \cdot 2.35, \ \hat{X} + (s/\sqrt{4}) \cdot 2.35]. \tag{1.5.2}$$

Your boss has doubts about the quality of your CI, because the normal assumption seems to him unrealistic. In order to persuade your boss that your proposal is reasonable, you decide to undertake Monte Carlo simulation for the case when you know in advance the mean value. As to the unknown distribution of X_i you decide to check two distributions which are quite "far" from normal: the Uniform $X \sim U[10,12]$, and the shifted Exponential $X \sim 10 + Exp(\lambda = 1)$. In both cases the population mean equals 11.

Below is the printout of *Mathematica* program for $X \sim U(10,12)$.

```
<< Statistics`ContinuousDistributions`
dist=UniformDistribution[0,2];
Q=100000:
sig=0;
For[i=1, i < Q + 1, i++,
x1=Random[dist]+10;
x2=Random[dist]+10;
x3=Random[dist]+10;
x4=Random[dist]+10;
xbar=(x1+x2+x3+x4)/4;
stdev=Sqroot((x1-xbar)²+(x2-xbar)²+(x3-xbar)²+(x4-xbar)²)/3);
left=xbar-2.35*stdev/2;
right=xbar+2.35*stdev/2;
alpha=Sign[(right-11)(11-left)]+1;
sig=sig+alpha/2];
conf=siq/Q; conf=N[conf];
Print["confidence=",conf]
```

Repeat the calculations following this program. The confidence level for $X \sim U(10,12)$ is about 0.88.

Modify the above program for the case of X having shifted exponential distribution with $\lambda = 1$.

2. Optimal Job-Shop Schedule.

Suppose you have n jobs, each given by a triple of numbers

$$Data(i) = [t_1(i), t_2(i), \Delta(i)], i = 1, ..., n, \tag{1.5.3}$$

where $t_1(i)$ is the preplanned beginning time, $t_2(i)$ is the preplanned completion time, and $\Delta(i)$ is tolerance on the beginning time, which is understood as follows:

the actual beginning time $T(i) \in [t_1(i) - \Delta(i), t_1(i) + \Delta(i)]$.

For sake of simplicity, all times are integers on a time grid with fixed step of say, 15 minutes. Note that the duration of the job remains constant for any permissible beginning time and equals $t_2(i) - t_1(i)$.

Each machine in the workshop can process at any moment in time one and only one work. Each work should be processed without interruptions.

Our goal is to find out the optimal beginning time for each job to *minimize the total number of machines* needed to process all jobs.

To formulate the main result needed to solve our problem we need to introduce a characteristic function $h_i(t)$ for job $i, i = 1, ...n$. If this job starts at time $T_1(i)$ and ends at $T_2(i)$, then

$$h_i(t) = 1, \text{for } t \in [T_1(i), T_2(i)], \text{ and } h_i(t) = 0, \text{ otherwise.} \tag{1.5.4}$$

Dilworth's Theorem [17].

The minimal number of machines to process all jobs with fixed beginning times is equal to the maximum of the function

$$D(t) = \sum_{i=1}^{n} h_i(t). \ \#. \tag{1.5.5}$$

We could find the optimal position for each job by a direct enumeration. This is not, however, always possible. For example, we have $n = 30$ jobs, and each tolerance equals ± 5. So the total number of variants is 11^{30}, a huge number. Another approach would be to try a random choice of the job positions for all jobs. This would work as follows: check the maximal value of the function $D(t)$; choose another random location of all jobs and stop after, say, 10,000 trials. Pick up the schedule which provided the *minimal* value of the maximum of $D(t)$.

Note that the number of machines cannot be smaller then some number $Dmin$ which is equal to the integer part $+1$ of the following ratio: total duration of all jobs divided by the length of the time span designated for processing all works. If our random search gives $\max D = Dmin$, we have achieved our goal.

This approach can be refined by means of the following heuristics. Start with a random location of jobs. Analyze the neighborhoods of the maxima points of the function $D(t)$. Usually there is a small number of jobs located near a maximum point. Try a total enumeration of all possibilities for locating these jobs. (Try only extreme cases for Δ's.) There are good chances that the maximal value of $D(t)$ will be lowered by at least one unit.

We suggest you design your own algorithm for finding the optimal schedule.

Let us outline another idea which was successfully implemented in the papers cited below. It is based on the notion of *entropy*.

For a discrete distribution $f = (f_1, f_2, ..., f_m)$ with probabilistic mass located in the points $1, 2, ..., m$, the entropy $E(f)$ is defined as follows:

$$E(f) = -\sum_{i=1}^{m} \log f(i) \cdot f(i). \tag{1.5.6}$$

The entropy has the following fundamental property: of all distributions with the same support, the **uniform** distribution has the largest entropy.

Now note that the above defined function $D(t)$ can be transformed into distribution by appropriate normalization, by dividing by some constant C. Suppose this has been done. Let us preserve the same notation for the normalized function $D(t)$. It is defined on a discrete grid and we can define its entropy $E(D)$ by the formula above. Maximal entropy would mean most uniform "spread" of the function $D(t)$, which means the <u>minimal</u> number of machines needed to process all jobs. This observation gives rise to the following idea first proposed by Kh. Kordonsky, [25,33].

Locate all jobs randomly within their tolerances. Compute the corresponding entropy. Repeat this operation K times and obtain the average entropy $\bar{D}$. Delete from the schedule job j, and repeat the same procedure for the set of all jobs without job j. Obtain the corresponding average entropy $\bar{E}(-j)$. Compute the difference in the entropies $\Delta(j) = \bar{D} - \bar{E}(-j)$.

The crucial idea of the proposed algorithm is the following: the first priority is given for the "most demanding" job $j^\star$ for which $\Delta(j)$ is **maximal**.

Locate this job randomly into the schedule, repeat the search for the next candidate for the remaining jobs, etc. The algorithm is of "greedy" type and the computer time to perform it is linear as a function of the number of jobs.

The paper [33] gives an example of scheduling 32 jobs, with total number of different locations $7.4 \cdot 10^{25}$. The optimal schedule needs 7 machines. The above described entropy priority rule has found, after 10,000 schedules were composed, the schedule which needs 8 machines. An improvement of the algorithm (not described here) has found the optimal solution for 7 machines. Interesting to note that the best of 10,000 schedules with random positioning of jobs within their tolerances and random priority rule was for 11 (!) machines.

As an exercise, design a simplified algorithm for finding an optimal location of 10 jobs given by the following triples $[b(i), c(i), \delta(i)], i = 1, ..., 10$, where i is the job number, $b(i)$ is its integer beginning time, $b(i) \in [1, 20]$, $c(i)$ is its completion time, $c(i) \in [1, 20], c(i) > b(i)$, and $\delta(i)$ is the tolerance on the beginning time meaning that the i-th job can start in the interval $b(i) - \delta(i), b(i) + \delta(i)$.

The idea of the algorithm is the following: for job i, generate an integer uniform random variable $X_i \sim$ **UINTEGER**$[-\delta(i), +\delta(i)]$; locate the actual job on the time scale with beginning time $b(i) + X_i$ and completion time $c(i) + X_i$. After doing it for all 10 jobs, calculate the number K of machines needed to perform all jobs, repeat the experiment 10,000 times, and remember the best schedule providing the minimal K. In fact this is a simplified version of the above described search procedure, which originally was based on using the entropy approach.

3. Lucky Numbers: a Toy Example.

We have decided to open a new business - a lottery. Each participant draws 6 random digits uniformly distributed on the integers $0, 1, 2, 3, ..., 9$. Afterwards, he (she) sums up all of them and receives from us \$1000 if the sum is a lucky number: $4, 9, 16, 25, 36$, or 49. (Only squares are lucky numbers.) Participation in one game costs \$120. Would this lottery be profitable for us?

Obviously, to answer this question it is necessary to know the probability to win in a single game, i.e. the probability to get a lucky number.

To find out this probability, we have decided to carry out a Monte Carlo experiment.

Try to do it as an exercise! Write a *Mathematica* program to find the probability to draw a lucky number in a single game.

4. Prove the formula $E[R] = N \sum_{k=1}^{n} a_k \cdot p_k$ for the mean reward in multinomial experiment.

5. In a multinomial experiment with n outcomes, the outcome Ω_i appears with probability f_i, $i > 2$, $f_1 + f_2 + ... + f_n = 1$. This experiment was carried out N times, the outcome Ω_i was observed N_i times, $N = \sum_{i=1}^{n} N_i$. Denote by $\hat{f}_i$ the relative frequency of the outcome Ω_i: $\hat{f}_i = N_i/N$. Define $R = \sum_{i=1}^{n} \hat{f}_i a_i$.
Derive the formula for $Var[R]$.
Hint. Follow the derivation given in Section 1.4 for $n = 3$.

6. *Continuation of Problem 5.* Prove that for $a_i > 0$,

$$r.e.[R] < \frac{(\max(f_1, ..., f_n))^{0.5}}{N^{0.5} \min(f_1, ..., f_n)}. \tag{1.5.7}$$

7. We say that random variable X has a *uniform distribution* on the interval $[0, 1]$ if the density function of X is $f_X(t) = 1$ for $t \in [0, 1]$ and zero, otherwise. We use the notation $X \sim U(0, 1)$.
a. Find the mean value μ of X, $\mu = E[X]$, the variance, $\sigma^2 = Var[X]$, the so-called coefficient of variation $c.v.[X] = \sigma/\mu$.
b. Find the CDF of X, $F_X(t) = P(X \leq t)$.

8. Continuation of 7. Let $X \sim U(0, 1)$. We say that random variable Y has a *uniform distribution* on the interval $[\alpha, \beta]$, $\beta > \alpha$ if the density function of Y, $f_Y(t) = 1/(\beta - \alpha)$ for $t \in [\alpha, \beta]$ and zero, otherwise. We denote this random variable as $Y \sim U(\alpha, \beta)$. Suppose $Z = C \cdot X + D$, $C > 0$. Find the density function of Z, the CDF of Z, $E[Z]$ and $Var[Z]$.

9. Suppose you have a random number generator producing replicas of $X \sim U(0, 1)$. Design a random number generator for obtaining r.v. $Y \sim U(-1, +1)$.

10. Suppose $X_1, X_2, ..., X_n$ are independent random variables, each uniformly distributed in $[0, 1]$. Define $Y = \max(X_1, X_2, ..., X_n)$. Find the CDF of Y and its mean value.

11. Let $X_i \sim B(1, p)$, $i = 1, ..., n$, X_i are independent r.v.'s. $X_1 = 1$ with probability p and zero, otherwise. We can say that X_i *counts* the number of successes in a single binomial experiment. It is easy to find out that $E[X_i] = p$, $Var[X_i] = p(1 - p)$. Define $Y = \sum_{1}^{n} X_i$. Y is the random number of successes in n binomial experiments. We denote $Y \sim B(n, p)$. This representation immediately gives that $E[Y] = np$, $Var[Y] = np(1 - p)$.

Prove that

$$P(Y = k) = \frac{n!}{k!(n-k)!} p^k (1-p)^{n-k}.$$

Chapter 2

What is Network Reliability?

> Networks are everywhere.
> Mark Newman

2.1 Introduction

2.1.1 General description

Networks is a vast field which deals with a wide spectrum of real-life systems in industry, communication, software engineering, etc. Below are some WEB definitions of various networks which demonstrate high importance and wide usability of networks in different areas.

⋆ An interconnected system of things or people;

⋆ A communication system consisting of a group of broadcasting stations that all transmit the same programs;

⋆ A system of intersecting lines or channels;

⋆ A system of interconnected electronic components or circuits;

⋆ A group of computers, connected by a telecommunication links, that share information;

⋆ A group of stations connected by communication facilities for exchanging information.

In a non formal language, a network is any system which can be thought of and graphically represented as a collection of small circles (nodes) interconnected by lines (edges).

In transportation, the circles might represent delivery points of goods, the lines are the roads between them. In an urban street map, the nodes are the intersections, the edges are the streets.

In communication, the nodes might be information transmitters/receivers and the edges - the telecommunication channels and the cables connecting them.

In computer networks, the nodes represent computers and the edges - the information transmission lines connecting these computers.

In social networks, nodes might represent individuals and the edges correspond to the mutual direct correspondence between them via, for example, the Internet.

2.1.2   Networks: Topology

Networks differ by their *topology*. Many network structures can be represented in a form of so-called *trees*. A tree is characterized by the fact that between any two nodes there is only a single path, see for example so-called bus network, Fig. 2.1(a). So-called star structures (Fig. 2.1(b)) are also trees. Very often, a tree has a recurrent structure, as shown on Fig. 2.1(c). A chain of military command from the upper level (headquarter) to lowest level - small field units - is represented by a tree, see Fig. 2.1(d), see also [27,28]. So-called *fault trees* appear in the reliability analysis in analyzing the causes of appearance of a dangerous event [2,15].

Computer networks have a large variety of topologies. If each computer (computation center) in the system is directly connected to each other, the network is represented by a dense mesh, so-called complete network, see Fig. 2.1(e). The number of edges in complete networks is $(n^2 - n)/2$, where n is the number of nodes. In order to reduce the number of edges, computer networks often are organized into special sparse structures which provide good performance of the network but demand considerably smaller number of edges.

So-called n-dimensional *cube* network, see Fig. 2.1(f), and Fig. 10.2, has 2^n nodes and $n2^{n-1}$ edges. If nodes are denoted by n-digit binary numbers, then edges exist between any pair of nodes differing by a single digit in their binary representation. This topology of n-cube network provides optimal packet routing, i.e. minimal number of steps needed to reach any collection of N packets originated at single nodes to any collection of destination nodes, see [39], Chapter 4. Another very popular sparse network structure is so-called *butterfly* shown on Fig.2.2(a). Butterfly of order $n = 4$ shown on Fig. 2.2(a) has $N = n \cdot 2^{n-1}$ nodes and $m = 2N$ edges. Nodes are organized in 2^{n-1} columns and n rows. In so-called *wrapped* butterfly, the nodes of the upper row are glued to the corresponding nodes in the first row. Wrapped

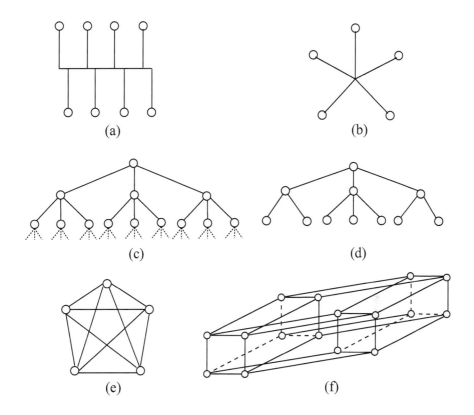

Figure 2.1: Various network topologies

butterfly has an important symmetry property that each node is incident exactly to four edges. Similarly to cube networks, butterfly networks have optimal property in packet transmission routing [39], Chapter 4.

In communication, a popular way of information transmission are so-called channel networks [41]. The nodes of this network are divided into k levels, the first level being a single source node s and the k-th -level - a single terminal node. Edges connect nodes only between adjacent levels, see Fig. 2.2(b).

It is impossible not to mention the Internet network. An excellent source on Internet properties is the book [14]. If we consider Internet as a very large social connection network, then one of its amazing properties is so-called small world property: the shortest path leading from any node to any other node is, on the average, very small, about 5-6. This is provided by the fact that the network node weights (measured by the number of edges incident to them) have so-called heavy tail distribution. Simply speaking, there is a

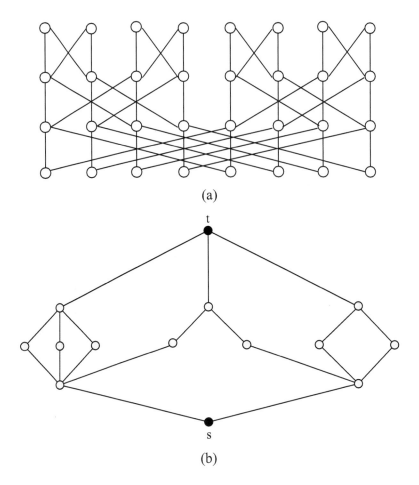

Figure 2.2: Butterfly network of order 4 (a); Channel network (b)

relatively small number of very heavy "popular" nodes with large number of edges and majority of nodes with relative small weight.

2.1.3   Networks: Reliability perspective

In different networks, we are interested in different performance parameters. For urban road networks, important points are car traffic characteristics, like waiting time, travel time index, congestion, etc. Transportation networks usually are studied to determine the maximum capacity flow between *source* node and *terminal* node and/or the characteristics of the shortest paths. For information transmission networks, important parameters are maximal transmission speed and information transmission capacity.

Reliability theory in general and this text in particular, studies networks mainly from the point of view of several of their principal *reliability indices*. Let us briefly describe them. *Nodes* and *edges* are the main network components (elements) which are subject to failure. *Edge* failure means that it does not exist, or is erased. If a *node* fails, then all edges incident to it are erased. Networks have several nodes which are of special importance and are assumed not to fail. They are called *terminals*. One of the most important network reliability parameters is so-called probability of terminal connectivity, i.e. the probability that all terminals are connected to each other. In particular, if all nodes are terminals, we speak about *all-terminal connectivity*. In many cases, there are only two terminals, and it is vital to provide the possibility to reach one of them from another one. This leads to probability of so-called *source-terminal* or $s - t$ *connectivity*. For example, in channel networks, the nodes on the lowest and the highest level are the terminals, see Fig. 2.2(b). The failure probability for this network is often called "blocking" probability, since the failure here means that each path from s to t is disrupted.

Valuable network reliability parameters are *component importance measures*. Component a importance measure characterizes the contribution to the network reliability which comes from improving component a reliability.

In this book we assume that each network element (node, or edge) can only be in two mutually exclusive states, *up* and *down*: *down* means failure, *up* - normal functioning.

It is important to understand that any statement of type

"Probability that component a is *up* equals p_a", can be interpreted in two ways, "*static*" and "*ergodic*", or "*stationary*". The *static* interpretation means that *time* is not present in our consideration. The state of element a is determined by means of a virtual binary statistical experiment with success probability p_a and failure probability $q_a = 1 - p_a$.

We use notation "*up*" and "*down*" to denote the element (component) state, and capital italic *UP* and *DOWN* to denote the whole system (network) state.

The outcomes of all element "lotteries" determine the (static) probability that the whole network is *UP* or is *DOWN*. For example, if network *UP* is defined as the overall connectivity, and the element lottery produces down results for all edges incident to a particular node a, then the network will be *DOWN*, since a becomes isolated. Similarly, if all paths leading from source s to terminal t are disrupted, the channel network is *DOWN*.

The *ergodic* or *stationary* interpretation means the following. It is assumed that each network component, independently of others, has random periods of being *up*, alternating with random periods of being *down*, e.g. for repair. Suppose that components functioning starts at time $t = 0$ and the mean *up* and *down* periods for component a are $\mu_a(up)$, and $\mu_a(down)$, respectively. Then, under quite general conditions, the probability that the component will be *up* in some remote instant t, formally as $t \to \infty$, will be

$$p_a(up) = \frac{\mu_a(up)}{\mu_a(up) + \mu_a(down)}.$$

So, we may treat all network components (nodes and/or edges) as undergoing alternating operation-repair processes and relate our probabilistic conclusions to some remote time instant. Similarly, we may assume that the system exits and works already for a long time, and the present time instant serves as the "remote time".

When all components alternate being *up* and *down*, the whole system also has some time interval of being *UP*, after which it enters its *DOWN* interval, etc. Similarly to component stationary probability to be *up*, we define the *system stationary* probability to be *UP*

$$P_{\mathbf{N}}(UP) = \frac{\mu_{\mathbf{N}}(UP)}{\mu_{\mathbf{N}}(UP) + \mu_{\mathbf{N}}(DOWN)}.$$

In connection with this framework, interesting and important network reliability parameters are network mean *UP* and mean *DOWN* periods,
$\mu_{\mathbf{N}}(UP)$, and $\mu_{\mathbf{N}}(DOWN)$.

Another view on network reliability will be obtained if we assume that each network component a has *random lifetime* $\tau_a \sim F_a(t)$. At time $t = 0$, all components start their lives and are up. At some random instant $\tau_{\mathbf{N}}$ the network goes *DOWN* and remains in this state. We will be interested in the distribution function of the network lifetime $F_{\mathbf{N}}(t) = P(\tau_{\mathbf{N}} \leq t)$. Knowing this function allows finding network reliability at any given instant t_0 as $R_{\mathbf{N}}(t_0) = 1 - F_{\mathbf{N}}(t_0)$, i.e. the probability that the network is *UP* at instant t_0.

2.2   Spanning Trees and Kruskal's Algorithm

2.2.1   Spanning tree: definitions, algorithms

This subsection is devoted to the description of so-called spanning trees of the network and to the algorithms for efficient construction of these trees.

Finding spanning trees (or checking whether such do exist) is the vital part of almost all algorithms for finding network reliability parameters described in this book. Let us present some basic definitions.

Definition 2.2.1. A directed graph (or digraph) G is a pair (V, E), where V is a finite set and E is a binary relation on V. The set V is called the *vertex* set of G, and its elements are called vertices (also nodes).

The set E is called the *edge set* of G, and its elements are called edges. In an undirected graph $G = (V, E)$, the edge set E consists of *unordered* pairs of vertices (nodes). By convention, we use the notation $e = (u, v)$ for an edge, where u and v are the appropriate nodes.

An undirected graph is connected if every pair of vertices is connected by a path. Figure 2.3 below represents undirected and directed graphs.

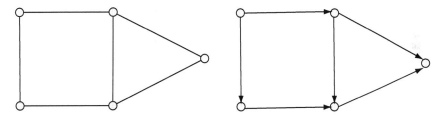

Figure 2.3: Undirected and directed graphs

Definition 2.2.2. A *tree* is an acyclic, connected, and undirected graph. Equivalently, a tree is an undirected graph in which there exists exactly one path between any given pair of nodes.

We present without proof the following important properties of trees:

(1) A tree with n nodes has exactly $(n-1)$ edges; (2) If a single edge is added to a tree, then the resulting graph contains exactly one cycle; (3) If a single edge is removed from a tree, then the resulting graph is no longer connected.

For example, edges (1,4), (4,3), (3,2), and (3,5) constitute a tree for the graph shown on Fig. 2.4(a). If we add one edge, e.g. (1,2), a cycle of nodes (1,4,2) will be created. Removing any edge from the tree, e.g. (4,2) disrupts the connection between nodes 1,4 and the remaining nodes 2,3,5.

Definition 2.2.3. Let $G = (V, E)$ be a connected and undirected graph. By Spanning Tree (ST) we call a subset T of E such that (V, T) is a tree and all nodes remain connected when only the edges in T are used.

Suppose now that each edge has a given *length*. When the sum of the lengths of the edges in T is as small as possible, we say that T is a Minimal

Spanning Tree (MST) of G. Similarly we can define a maximal spanning tree (MaxST) of G which is a spanning tree with maximal sum of its edge lengths. Note that it is possible that a graph has many MST's (and MaxST's).

As was mentioned in the beginning of this subsection, the notion of ST is of great importance to many reliability problems related to networks. Almost all algorithms in our book will be dealing with constructing spanning trees.

Example 2.2.1. In Fig. 2.4 we see a graph with given edge lengths and one of the MST's of this graph. The tree (1,4), (4,2), (2,3), (3,5) is a spanning tree, but not an MST. An MST is presented on Fig. 2.4(c).

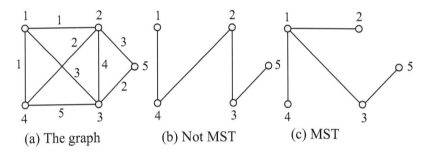

Figure 2.4: The graph and its MST

The most famous classic algorithms for constructing MST's are the Kruskal's and Prim's algorithms. Both algorithms are very simple (but it does not mean that proving their correctness is also very simple!). The idea of these algorithms is as follows. In Kruskal's algorithm, we start with empty set T of edges and at every stage we select the shortest edge that has not yet been chosen, so that adding this edge to T does not create a cycle. In the Prim's algorithm, we chose initially some node and on every stage we construct a tree from there, by selecting the shortest available edge that can extend the tree to an additional node. It may seem that it makes no difference which of the two algorithms we use, but it is not so. It may be proved that using Kruskal's algorithm is more efficient for the case of relatively small number of edges (so-called *sparse* networks), whereas the Prim's algorithm is more efficient for the case of relatively large number of edges (so-called *dense* networks). In this book we will deal with the Kruskal's algorithm.

The scheme of Kruskal's algorithm can be represented in the following simple form:

Algorithm 2.1 - Scheme of Kruskal's Algorithm.

1. **Sort** all edges by their length in increasing order.

   Denote the spanning tree by ST.

2. **Let** index $i := 0$. **Let** $ST := \{\}$.

3. **Choose** the next edge (a, b) from E.

4. **If** adding this edge to ST creates a cycle, **GOTO** 3.

5. $ST := ST \cup \{(a, b)\}$.

6. $i := i + 1$.

7. **If** $i < n$ **Then GOTO** 3. Else **STOP**.

Example 2.2.2. Consider the graph in Fig. 2.4. Suppose that after sorting we get the edges in the following order: $(1, 2), (1, 4), (2, 4), (3, 5), (2, 5), (1, 3),$ $(2, 3), (3, 4)$.

Now the algorithm proceeds as follows.

Step 1. $i = 0, ST = \{\}$.

Step 2. Add edge(1,2). $ST = \{(1, 2)\}, i = 1$.

Step 3. Add edge (1,4). $ST = \{(1, 2), (1, 4)\}, i = 2$.

Step 4. Reject edge (2,4) since adding it creates a cycle.

Step 5. Add edge (3,5). $ST = \{(1, 2), (1, 4), (3, 5)\}, i = 3$.

Step 6. Add edge (1,3). $ST = \{(1, 2), (1, 4), (3, 5), (1, 3)\}, i = 4$.

The MST is constructed and is given by the set ST obtained in Step 6. It is clear from the above scheme and the example that the main difficulty lies in checking the possibility of a cycle appearance. This can be done by using so-called *Disjoint Set Structures* (DSS). We describe this important instrument in the following subsection. We would like to note that there exist different efficient schemes to implement the Kruskal's algorithm. Our criterion in choosing and describing the appropriate scheme was simplicity, so that every reader may turn the pseudocodes of the next subsection into a working computer program. Of course, an experienced reader may improve the proposed pseudocodes and write his/her own more efficient programs.

In conclusion, a remark about constructing the MaxST. All the above described algorithms remain exactly the same, with replacing the choice of the *minimal* length (weight) edge by *maximal* length (weight) edge. So, in Kruskal's algorithm we choose on each step the longest edge from the remaining edges.

2.2.2   DSS - disjoint set structures

Suppose we have n objects numbered from 1 to n (in our case we can think about these objects as network nodes). We will need to group these objects into disjoint sets (components), so that at any given stage each object is in exactly one set. Initially, all objects are in n different sets, each containing exactly one object. We will need to execute a sequence of operations of two kinds:

1. Given some object, we find out which set contains it, and return the label of this set (component).

2. Given two different labels, we *merge* the contents of the two corresponding sets, and choose a label for the combined set.

These two operations are the main parts of the Kruskal's algorithm. Indeed, when we choose the next edge $e = (v, w)$ we check to which components belong its incident nodes v and w. In cases where they belong to the same component, a cycle has been created, and we reject this edge. In cases where the components are different, we add this edge. It is worth mentioning that the efficient implementation of the algorithm is of great importance. To demonstrate this we present here two implementations of the operations on DSS: one straight and "naive" and the other - much more efficient.

The straightforward implementation is given by the two following pseudocodes.

Suppose that an array $Comp[1, ..., n]$ is given. For each object i, $Comp[i]$ means the label of the set containing i.

Function $find1(x)$
   return $Comp[x]$

In words: this function receives the object (node) x and returns the appropriate component $Comp[x]$. The following procedure merges the sets labeled a and b (assume that $a \neq b$).

Procedure $merge1(a, b)$
   for $k \leftarrow 1$ **to** n
      if $Comp[k] = b$ **then** $Comp[k] \leftarrow a$.

In words: this procedure receives two components labeled a and b and produces the combined component labeled a. Note that in the latter procedure it makes no difference, a or b is chosen as a label for the resulting set. Suppose now that we construct the MST by Kruskal's algorithm. Then we need at least $2 \cdot (n - 1)$ calls to $find1$ and $(n - 1)$ calls to $merge1$. Taking into

account that *merge*1 contains a loop of size n we arrive at the conclusion that the implementation of Kruskal's algorithm takes a time of $O(n^2)$. We will describe now a more efficient version. It is based on two principal ideas:
(a) tree representation for disjoint sets.
(b) more efficient criterion for label choice in sets merging.

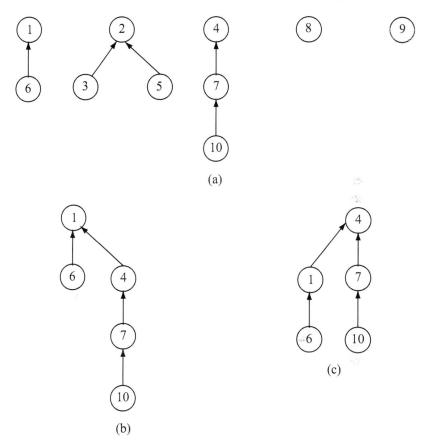

Figure 2.5: (a) The tree representation of DSS; (b) "Sticking" $(10 \rightarrow 7 \rightarrow 4)$ to $(6 \rightarrow 1)$; (c) "Sticking" $(6 \rightarrow 1)$ to $(10 \rightarrow 7 \rightarrow 4)$

By (a) we represent each set (component) as a rooted tree, where each node contains a reference to its parent. We adopt the following scheme: if $Comp[i] = i$, then i is both the label of its set and the root of the corresponding tree; if $Comp[i] = j \neq i$, then j is the parent in some tree.

Example 2.2.3. Consider the following array: $\{1, 2, 2, 4, 2, 1, 4, 8, 9, 7\}$. It represents the following disjoint sets: $\{1,6\}, \{2,3,5\}, \{4,7,10\}, \{8\}, \{9\}$. In Fig. 2.5 we can see the corresponding tree representation for these sets. Now finding the set label for an element means "jumping" from a node to a node till the condition $set[i] = i$ will be fulfilled. For example, from the given array (and also from the Figure 2.5(a)) we see that $4 = find(10)$.

The second idea is based on the following simple fact. Suppose that the tree tr_1 is shorter than tr_2. Then "sticking" tr_1 to tr_2 gives a shorter tree (Fig. 2.5(b)) than "sticking" in opposite order, see Fig. 2.5(c).

The above simple ideas allow to realize the operations $find$ and $merge$ in a more efficient way. The appropriate function $find2$ is given by the following pseudocode.

Function $find2(x)$

   $r \leftarrow x$

   while $Comp[r] \neq r$

      $r \leftarrow Comp[r]$

   return r

For the new pseudocode $merge2$ we need to define array $H[1, ..., n]$, which presents the heights of the appropriate subtrees, so that $H[i]$ is the height of the subtree presenting the set labeled by i. For example, for the trees on Fig. 2.5(a) we have: $H[2] = 1$, $H[4] = 2$. Initially (i.e when each set $Comp[i], i = 1, ..., n$ is consisting of one element), we have $H[i] = 0, i = 1, ..., n$.

Function $merge2(a, b)$

   if $H[a] = H[b]$

   then

      $Comp[a] \leftarrow Comp[b]$

      $H[b] \leftarrow H[b] + 1$

      return b

   else

      if $H[a] < H[b]$

      then

         $Comp[a] \leftarrow Comp[b]$

         return b

      else

         $Comp[b] \leftarrow Comp[a]$

         return a

For example, after applying $merge2(1,4)$ we get the tree on Fig. 2.5(b) with $H[4] = 2$. It may be proved (see [5]) that by using $merge2$ we get trees with the height at most $\lceil \log_2(n) \rceil$. It follows that the operation $find2$ needs time $O(\log_2(n))$, i.e. using $find2$ and $merge2$ is much more effective than using $find1$ and $merge1$.

Now we are ready to present the pseudocode for the Kruskal's algorithm, for both cases of connectivity - the k-terminal connectivity and the overall connectivity. The pseudocode uses the following arrays.

a. $Edge[1, ..., m]$ - the array of edges sorted by their lengths, so that $Edge[i]$ is the number of the appropriate edge.

b. $Fnode[1, ..., m]$ - the array of edge first nodes.

c. $Snode[1, ..., m]$ - the array of the second edge nodes. So the edge $Edge[i]$ has two nodes: $Fnode[i]$ and $Snode[i]$. We suppose that $Fnode[i] < Snode[i]$.

d. $Comp[1, ..., n]$ - the array of components for nodes, so that the node i belongs to the component $Comp[i]$.

e. $T[n_1, ..., n_k]$ - the array of terminal numbers.

f. $ST[1, ..., n-1]$ - the array of the spanning tree.

g. $Tcomp[1, ..., n]$; $Tcomp[i]$ gives the number of terminals belonging to the component i.

h. $H[1, ..., n]$ - the array defined above for $merge2$.

Procedure Kruskal

// Initialization

```
for i ← 1 to n
H[i] ← 0
Comp[i] ← i
Tcomp[i] ← 0
for i ← 1 to k
    j ← T[i]
    Tcomp[j] ← 1
```

// $iTcomp$ - the current maximal number
// of terminals in one component.

```
iTcomp ← 1
```

// iST - the number of edges in
// the current state of the spanning tree.

```
iST ← 0
i ← 0
```

repeat
$\qquad i \leftarrow i + 1$
$\qquad u \leftarrow Fnode[i]$
$\qquad v \leftarrow Snode[i]$
$\qquad ucomp \leftarrow find2(u)$
$\qquad vcomp \leftarrow find2(v)$
$\qquad\qquad\qquad\qquad$ // check if the nodes of the edge i
$\qquad\qquad\qquad\qquad$ // belong to different components.
$\qquad$ **if** $(ucomp \neq vcomp)$
$\qquad$ **then**
$\qquad\qquad r \leftarrow merge2(ucomp, vcomp)$
$\qquad\qquad iST \leftarrow iST + 1$
$\qquad\qquad ST[iST] \leftarrow Edge[i]$
$\qquad\qquad\qquad\qquad$ // the number of terminals in the
$\qquad\qquad\qquad\qquad$ // resulting component after merging.
$\qquad\qquad j \leftarrow Tcomp[ucomp] + Tcomp[vcomp]$
$\qquad\qquad Tcomp[r] \leftarrow j$
$\qquad\qquad$ **if** $(j > iTcomp)$
$\qquad\qquad$ **then**
$\qquad\qquad\qquad iTcomp \leftarrow j$
$\qquad$ **until** $(iTcomp = k)$

Example 2.2.4.

Let us implement all stages of the Kruskal procedure for the network with 3 terminals shown on Fig. 2.6. Table 2.1 presents the sorted edges given together with their incident nodes.

Initially we have the following values for the arrays and variables. $T = \{1, 4, 6\}$, $iTcomp = 1$, $iST = 0$, $Comp = \{1, 2, 3, 4, 5, 6, 7\}$, $Tcomp = \{1, 0, 0, 1, 0, 1, 0\}$, $H = \{0, 0, 0, 0, 0, 0, 0\}$.
Step 1. $u = Fnode(2) = 2$, $v = Snode(2) = 3$, $ucomp = 2$, $vcomp = 3$, $r = merge2(2, 3) = 3$, $Comp = \{1, 3, 3, 4, 5, 6, 7\}$, $Tcomp = \{1, 0, 0, 1, 0, 1, 0\}$, $H = \{0, 0, 1, 0, 0, 0, 0\}$, $iTcomp = 1$, $ST = \{2\}$.
Step 2. $u = Fnode(7) = 1$, $v = Snode(7) = 7$, $ucomp = 1$, $vcomp = 7$, $r = merge2(1, 7) = 7$, $Comp = \{7, 3, 3, 4, 5, 3, 7\}$, $Tcomp = \{1, 0, 0, 1, 0, 1, 1\}$, $H = \{0, 0, 1, 0, 0, 0, 1\}$, $iTcomp = 1$, $ST = \{2, 7\}$.
Step 3. $u = Fnode(9) = 2$, $v = Snode(9) = 6$, $ucomp = 3$, $vcomp = 6$, $r = merge2(3, 6) = 3$, $Comp = \{7, 3, 3, 4, 5, 6, 7\}$, $Tcomp = \{1, 0, 1, 1, 0, 1, 1\}$, $H = \{0, 0, 1, 0, 0, 0, 1\}$, $iTcomp = 1$, $ST = \{2, 7, 9\}$.

Table 2.1: The Sorted Edges and Their Nodes

Edge	Fnode	Snode
2	2	3
7	1	7
9	2	6
10	3	6
8	2	7
6	6	7
1	1	2
3	3	4
4	4	5
11	3	5
5	5	6

Step 4. $u = Fnode(10) = 3$, $v = Snode(10) = 6$, $ucomp = 3$, $vcomp = 3$. Reject the edge.

Step 5. $u = Fnode(8) = 2$, $v = Snode(8) = 7$, $ucomp = 3$, $vcomp = 7$, $r = merge2(3,7) = 7$, $Comp = \{7,3,7,4,5,6,7\}$, $Tcomp = \{1,0,1,1,0,1,2\}$, $H = \{0,0,1,0,0,0,2\}$, $iTcomp = 2$, $ST = \{2,7,9,8\}$.

Step 6. $u = Fnode(6) = 6$, $v = Snode(6) = 7$, $ucomp = 7$, $vcomp = 7$. Reject the edge.

Step 7. $u = Fnode(1) = 1$, $v = Snode(1) = 2$, $ucomp = 7$, $vcomp = 7$. Reject the edge.

Step 8. $u = Fnode(3) = 3$, $v = Snode(3) = 4$, $ucomp = 7$, $vcomp = 4$, $r = merge2(7,4) = 7$, $Comp = \{7,3,7,7,5,6,7\}$, $Tcomp = \{1,0,1,1,0,1,3\}$, $H = \{0,0,1,0,0,0,2\}$, $iTcomp = 3$, $ST = \{2,7,9,8,3\}$. Stop the process.

As it was mentioned above, the Kruskal's algorithm is very useful in many applications. For example, if we want to check whether some network is connected, i.e. whether there is a path between each pair of nodes, then it is sufficient to construct a spanning tree with edges of arbitrary length. If we obtain that all nodes of the network belong to one component then the network is connected.

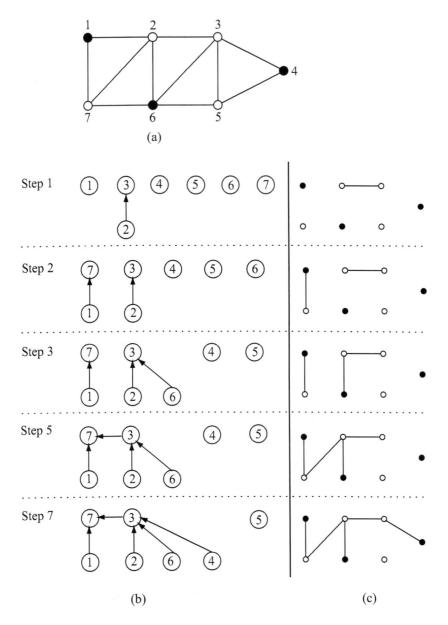

Figure 2.6: The graph (a) and Tree representation of DSS (b) and Steps of Kruskal's algorithm (c)

2.3   Introduction to Network Reliability

2.3.1   Static networks

In 2.1.3 the basic network reliability notions were presented. In this subsection we describe the elementary methods of reliability computing. Before we start with this, let us remind very briefly the following notions.

1. By network $\mathbf{N} = (V, E, T)$ we denote an undirected graph with a node-set V, $|V| = n$, an edge-set E, $|E| = m$, and a set $T \subseteq V$ of special nodes called *terminals*. For more detailed description see [5, 8].

2. Each element a (node or/and edge) is associated with a probability p_a of being *up* and probability $q_a = 1 - p_a$ of being *down*. We postulate that element failures are mutually independent events.

3. In a given network $\mathbf{N}$ the state of $\mathbf{N}$ is induced by all its elements which are in the *up* state.

4. In this book we deal with *terminal connectivity* operational criterion. By this criterion the state is UP if any pair of terminals is connected by the elements in the *up* state. In the case that the terminal set contains k terminals, $k < n$, we use the term *k-connectivity*. In the case of $T = V$, we will use the term *overall connectivity*. If $T = \{s, t\}$ we use the term *s-t connectivity*. The terminal connectivity has the property of being monotone: each subset of the $DOWN$ state is a $DOWN$ state, and each superset of the UP state is an UP state.

Let us now describe two elementary methods of computing the network reliability.

Straightforward computation. Suppose that we know the set S of **all** UP states of the network: $S = \{S_i, i = 1, ...r\}$. By our convention, S_i is a particular set of all network elements which are *up*, such that the network itself is UP. For simplicity, assume that only edges are subject to failures. Then, denoting by Q_i the complement of S_i,

$$P(S_i) = \prod_{e \in S_i} p_e \prod_{e \in Q_i} q_e, \qquad (2.3.1)$$

and

$$R(\mathbf{N}) = P(\mathbf{N} \ is \ UP) = \sum_{i=1}^{r} P(S_i). \qquad (2.3.2)$$

Example 2.3.1.

(a) Suppose we have a network with three nodes $V = \{1, 2, 3\}$ and three edges $e_1 = (1, 2), e_2 = (1, 3), e_3 = (2, 3)$, see Fig 2.7. The nodes are absolutely reliable. The operational criterion is overall connectivity. Edge e_i is *up* with probability p_i, $i = 1, 2, 3$. Edges are independent.

It is easy to check that there are four UP states: $S_1 = \{e_1, e_2\}, S_2 = $

$\{e_1, e_3\}, S_3 = \{e_2, e_3\}, S_4 = \{e_1, e_2, e_3\}$. Then by (2.3.1) and (2.3.2),

$$R(\mathbf{N}) = \sum_{i=1}^{4} P(S_i) = p_1 p_2 q_3 + p_1 p_3 q_2 + p_2 p_3 q_1 + p_1 p_2 p_3. \qquad (2.3.3)$$

(b) Suppose now that $T = \{1, 2\}$, i.e. nodes 1 and 2 are terminals, node 3 remains reliable. For this case, the set of all UP states is $S = \{S_1 = \{e_1\}, S_2 = \{e_1, e_2\}, S_3 = \{e_1, e_3\}, S_4 = \{e_2, e_3\}, S_5 = E\}$. In this case we have $R(\mathbf{N}) = p_1 q_2 q_3 + p_1 p_2 q_3 + p_1 p_3 q_2 + p_2 p_3 q_1 + p_1 p_2 p_3$.

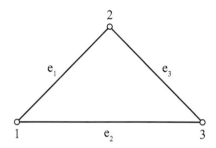

Figure 2.7: The network of Example 2.3.1

We can get from this example an impression that reliability computations are quite simple. The problem with the method of this example is that we have to enumerate all network states and check which of them are UP. Let us take a complete network of size 10, which is a network with 10 nodes and 45 edges connecting every pair of nodes. Suppose that only edges can fail. The number of all states equals $2^{45} = 35184372088832$, a huge number. Thus checking all states even for small networks is practically impossible.

Paths and Cuts. Let us define paths and cuts for networks with unreliable edges and reliable nodes.

Definition 2.3.1. A set $L = \{e_1, e_2, ..., e_r\}$ is called a *path* if the network is in an UP state if *all* elements of the path are *up*. We say that the set L is a *minimal* path if there does not exist such an element e^* so that $L^* = L - \{e^*\}$ remains a path.

For example, for all-terminal connectivity, a spanning tree is a path. Moreover, it is a minimal path, since as we proved earlier, a spanning tree has no redundant edges. For the network on Fig. 2.7 edges $L_1 = \{e_1, e_2, e_3\}$ constitute a path, but not a minimal path because any edge can be deleted from L_1 without destroying connectivity.

Definition 2.3.2. A set $C = \{e_1, e_2, ..., e_l\}$ is called a *cut*, if the network is *DOWN* if *all* elements of C are *down*. We say that C is a *minimal* cut if there does not exist an element in C so that $C^* = C - \{e\}$ remains a cut. In words: a minimal cut has no redundant elements.

Minimal Path-set Method. Denote by $S_L = \{L_1, L_2, ..., L_r\}$ the set of all minimal paths of the network.

Then the following formula is valid:

$$R(\mathbf{N}) = P\left(\bigcup_{i=1}^{r} L_i^{up}\right), \tag{2.3.4}$$

where L_i^{up} stands for the following event: all elements in L_i are *up*. Indeed, it is easy to see that each UP state contains some minimal path. For example, in 2.3.1(b) the UP state S_5 contains two minimal paths: S_1 and S_4. On the other hand, each minimal path defines some UP state. So, the sufficient and necessary condition for $\mathbf{N}$ to be in the UP is that in at least one minimal path all elements are *up*.

Minimal Cut-set Method. Denote by $S_C = \{C_1, C_2, ..., C_l\}$ the set of all minimal cuts of a network.

Then the following formula is valid:

$$1 - R(\mathbf{N}) = P\left(\bigcup_{i=1}^{l} C_i^{down}\right), \tag{2.3.5}$$

where C_i^{down} stands for the following event: all elements in C_i are *down*. The proof of the latter equation is similar to the proof of (2.3.4).

Example 2.3.1 - continued. Let us consider again a three-node network which fails if all-terminal connectivity is violated. There are three min-path sets: $L_1 = (e_1, e_2), L_2 = (e_1, e_3), L_3 = (e_2, e_3)$. These sets are the spanning trees of the network. Using (2.3.4) we have to remember that the probability of a union of events must be computed by the inclusion-exclusion formula familiar to the readers from Probability Theory. So,

$P(L_1^{up} \cup L_2^{up} \cup L_3^{up}) = P(L_1^{up}) + P(L_2^{up}) + P(L_3^{up}) - P(L_1^{up} \cap L_3^{up}) - P(L_2^{up} \cap L_3^{up}) - P(L_2^{up} \cap L_3^{up}) + P(L_1^{up} \cap L_2^{up} \cap L_3^{up})$.

As an exercise, complete the calculations and check that you obtain the same result as in Example 2.3.1, (a).

The straightforward use of min-path or min-cut methods is generally not an efficient tool of computing network reliability, except for very small networks or networks with special structure, so-called series-parallel systems. We will deal with them in the following chapters.

2.3.2 Dynamic networks

In static networks, the state of network component (edge or node) is fixed once and forever on the basis of random lotteries. For example, e is up with probability $p(e)$ and down with probability $1 - p(e)$. The *time* coordinate is not present at all in these lotteries.

In dynamic networks, each edge or node has a *random lifetime*, which means the following. At the instant $t = 0$ all edges (or nodes, or both) are down. There exist a nonnegative random variable $\tau_e \sim F_e(t)$ which is the edge e "birth" time. So, with probability $F_e(\theta)$ the edge e will be "born", i.e. become up, before time θ (and afterward remain in this state forever). The lifetimes for different elements of the network are assumed to be *independent* random variables.

Very often we consider networks in which initially, at $t = 0$, all elements (nodes, edges or both) are *up*. Edge e remains *up* for a random time τ_e and afterwards goes down (and remains down "forever"). In other words, each component is equipped with a random lifetime τ. If $\tau = x$, at the instant $t = x$, this component becomes down and remains in this state.

The main problem for dynamic networks is finding the probabilistic description of the instant at which the whole network enters its $DOWN$ state (if it was initially UP), or becomes UP, if it was initially $DOWN$. Further in our course, in Chapters 8 and 9, we will study the dynamic networks and develop special techniques for evaluation network lifetime parameters.

2.4 Multistate Networks

In this subsection we present a very short description of so-called *multistate networks*. This type of networks is not at the center of our exposition in this book. Nevertheless, we think that for better understanding of network reliability concepts it is useful to have a more general picture of various types of networks. The reader interested in closer acquaintance with multi-state system reliability is referred to fundamental work [35].

Let us consider a network in which its elements have *three* states. One of these states is the *up* state. Two other states are *down* states. One of these *down* states is called *open* and another is called *short*.

The terminology *short* and *open* is borrowed from considering a diode: in the *up* state the diode allows passing of electric flow in only one direction. The *down* states correspond to two mutually exclusive situations: the electric flow passes on both directions (*short*) and when the electric flow does not pass at all (*open*).

As an example consider a *series* system of two elements and compute it reliability. Denote by $A_i, A_{i,o}, A_{i,s}$ the states *up*, *open*, and *short*, for element $i = 1, 2$, respectively. By the definition, the system can be in three states: one UP and two $DOWN$ states. The $DOWN$ state is defined as follows: at least one element is *open* or both elements are *short*.

Therefore, system reliability R equals

$$R = P(UP) = 1 - \{P(A_{1,o} \cup A_{2,o}) \cup (A_{1,s} \cap A_{2,s})\} \qquad (2.4.1)$$
$$= 1 - P(A_{1,o} \cup A_{2,o}) - P(A_{1,s} \cap A_{2,s}).$$

Assuming that the elements are independent and denoting $p_{i,o} = P(A_{i,o})$, $p_{i,s} = P(A_{i,s}), i = 1, 2$, we obtain

$$R = (1 - p_{1,o}) \cdot (1 - p_{2,o}) - p_{1,s} \cdot p_{2,s}. \qquad (2.4.2)$$

A *parallel* system of two elements, by definition, is in $DOWN$ state if one of the elements is in the *short* state or if both elements are in the *open* state. We arrive at the following expression:

$$R = P(UP) = 1 - \{P(A_{1,s} \cup A_{2,s}) \cup (A_{1,o} \cap A_{2,o})\} \qquad (2.4.3)$$
$$= 1 - P(A_{1,s} \cup A_{2,s}) - P(A_{1,o} \cap A_{2,o}).$$

Assuming that the elements are independent, we obtain

$$R = (1 - p_{1,s}) \cdot (1 - p_{2,s}) - p_{1,o} \cdot p_{2,o}. \qquad (2.4.4)$$

The reliability formulas can be easily generalized to the series and parallel systems of n elements.

We present without proof the following formulas for reliability of series and parallel systems, respectively:

$$R_{ser} = \prod_{i=1}^{n}(1 - p_{i,o}) - \prod_{i=1}^{n} p_{i,s}, \qquad (2.4.5)$$

$$R_{par} = \prod_{i=1}^{n}(1 - p_{i,s}) - \prod_{i=1}^{n} p_{i,o}, \qquad (2.4.6)$$

2.5   Network Reliability Bounds

Series, parallel, and combinations of series and parallel systems are not very often encountered in practice in "pure" form. There is however an interesting possibility to obtain *lower* and *upper* bounds on arbitrary monotone system reliability using the patterns of series and parallel connection of components. Let $A_1, A_2, ..., A_l$ be the complete list of all minimal path sets, and let $C_1, C_2, ..., C_m$ be the complete list of all minimal cut sets of a monotone system.

The following theorem was proved by Barlow and Proschan [2], Chapter 2:

Theorem 2.5.1. Denote by p_i the reliability of i-th component and by R_0 the system reliability. If all system components are statistically independent, then

$$\prod_{k=1}^{l}(1 - \prod_{j \in C_k}(1 - p_j)) \leq R_0 \leq 1 - \prod_{s=1}^{r}(1 - \prod_{j \in A_s} p_j). \qquad (2.5.1)$$

The theorem says that R_0 is bounded from above by the reliability of a fictitious system (with *independent* components) which is a *parallel* connection of series subsystems each being the minimal path set of the original system. Similarly, the lower bound is the reliability of a system obtained by a series connection of parallel subsystems each being the minimal cut set of the original system.

Fig. 2.8 illustrates this theorem.

We would like to stress that the components with identical numbers appearing in the lower and upper bounds, see Fig. 2.8, are considered as *independent* and *different* components. For example, components with number 1 appear in two minimal cuts. It is meant that we have a specially designed "lower bound" system with two *different* and *independent* copies of component 1 having *up* probabilities equal to *up* probability of component 1.

We omit the proof of the above important theorem. It is worth noting that the Barlow and Proschan's theorem is true not only for independent components but for dependent, or so-called *associated* components. For details see [2].

The method of network reliability estimation suggested by the above theorem seems very promising but in practice it is limited to small networks which allow complete enumeration of all their minimal cut sets and minimal path sets. In the case of the bridge network, it is easy to compose the list of

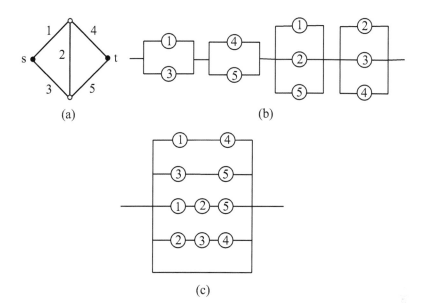

Figure 2.8: The bridge network (a) and the networks serving as its lower reliability bound (b) and upper reliability bound (c)

all minimal cuts and all minimal paths. The minimal cuts disrupt the $s-t$ connection and they are: $\{1,3\}, \{4,5\}, \{1,2,5\}, \{3,4,2\}$. The minimal paths providing the connection between s and t are: $\{1,4\}, \{3,5\}, \{1,2,5\}, \{3,2,4\}$. (The word "minimal" means that these sets do not contain redundant elements. For example, if we remove from the minimal path any of its elements, the new set will no longer be a path.)

2.6   Problems and Exercises

1. Complete the calculations of Example 2.3.1-continued in Section 2.3.

2. Compose the list of all minimal cuts containing two or three edges for the network shown on Fig. 2.9 below.

3. Suppose that for the given network $G = (V, E)$ all nodes are terminals. Suppose that p_e is the probability that edge e is *up*. Denote by ST a spanning tree of the network. The probability that the tree is UP is defined as

$$P(ST) = \prod_{e \in ST} p_e. \qquad (2.5.1)$$

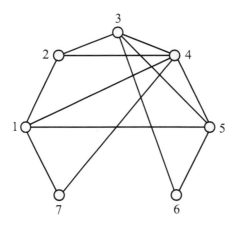

Figure 2.9: The network for Exercise 2. V=T, nodes are reliable

Describe an algorithm for finding the most reliable spanning tree for the given network.

4. For the bridge network on Fig. 2.8, compute the lower and upper reliability bounds assuming that all edges have probability p of being in the *up* state. Compare the upper and lower bound for $p = 0.9, 0.95, 0.99$.

5. Prove (2.3.4), (2.3.5).

Several problems below will be devoted to the reliability of so-called monotone binary systems. We introduce some new definitions.

We define a *system* (not necessarily a network) as a set of *components* (elements). For example, in a network, components may be edges and nodes. We assume that the components are *binary*, i.e. each component has only two states: operational (*up*) and failed (*down*). The state of component i will be described by a binary variable x_i, $i = 1, 2, ..., n$: $x_i = 1$ if the component i is *up* and $x_i = 0$ if the components i is *down*.

The whole system can only be in one of two states: *UP* and *DOWN*. The dependence of a system's state on the state of its components will be determined by means of so-called *structure function* $\phi(\mathbf{x})$, where $\mathbf{x} = (x_1, x_2, ..., x_n)$: $\phi(\mathbf{x}) = 1$ if the system is *UP* and $\phi(\mathbf{x}) = 0$ if the system is *DOWN*.

6. A system is called *series* system if it is *UP* if and only if all its components are *up*. A system is called *parallel* if it is *DOWN* if and only if all its components are *down*. Find the structure functions for series and parallel systems.

7. Consider the network on Fig. 2.7. Its components are the edges e_1, e_2, e_3. By definition, this network is UP if and only if all nodes are connected. Prove that

$$\phi(\mathbf{x}) = 1 - (1 - x_1 x_2)(1 - x_1 x_3)(1 - x_2 x_3).$$

8. Monotone systems. A system with structure function $\phi(\cdot)$ is called *monotone* if $\phi(\mathbf{x})$ has the following properties:
   (i) $\phi(0, 0, ..., 0) = 0$, $\phi(1, 1, ..., 1) = 1$
   (ii) $\mathbf{x} < \mathbf{y}$ implies that $\phi(\mathbf{x}) \le \phi(\mathbf{y})$.
   Explain the physical meaning of these properties.

9. Minimal paths. A state vector $\mathbf{x}$ is called a *path vector* if $\phi(\mathbf{x}) = 1$. The set of all *up* components of this vector is called **path set**. If, in addition, this path set does not contain redundant components, it is called *minimal path set*. Find all minimal path sets for the bridge system shown on Fig. 2.8(a). Is the set (1,2,3,4) a minimal path set?

10. *Structure function representation via minimal path sets.* Let $L_1, L_2, ..., L_r$ be the collection of all minimal path sets of the system. Prove that

$$\phi(\mathbf{x}) = 1 - \prod_{j=1}^{r}(1 - \prod_{i \in L_j} x_i). \tag{2.5.2}$$

In words: system structure function can be represented as a parallel connection of series systems representing minimal path sets.

11. Minimal cuts. A state vector $\mathbf{x}$ is called a *cut vector* if $\phi(\mathbf{x}) = 0$. The set of all down components of this vector is called **cut set**. If, in addition, for any $\mathbf{y} > \mathbf{x}$ $\phi(\mathbf{y}) = 1$, then the corresponding cut set is called *minimal cut set* or simply minimal cut. Find all minimal cut sets for the bridge system shown on Fig. 2.8(a). Is the set (1,2,3) a minimal cut set?

12. *Structure function representation via minimal cut sets.* Let $C_1, C_2, ..., C_l$ be the collection of all minimal cut sets of the system. Then

$$\phi(\mathbf{x}) = \prod_{j=1}^{l}[1 - \prod_{i \in C_j}(1 - x_i)]. \tag{2.5.3}$$

What is the physical meaning of this expression ?

13. Suppose a system has four minimal cut sets $(1, 2), (1, 5), (3, 4), (2, 5)$. Find all minimal path sets.
Hint. Represent the system in a form similar to Fig. 2.8(b), i.e. as a series connection of parallel subsystems, each made of the minimal cut.

For the next several problems we need some new definitions. Assume that the state of component i, $i = 1, 2, ..., n$ is described by a binary *random variable* X_i, where

$$P(X_i = 1) = p_i, \quad P(X_i = 0) = 1 - p_i, \tag{2.5.4}$$

where 1 and 0 correspond to the *up* and *down* state, respectively. The system *state vector* now becomes a random vector $\mathbf{X} = (X_1, X_2, ..., X_n)$. We assume that the r.v.'s X_i are independent. If $\phi(\mathbf{X}) = 1$, the system is *UP*. If $\phi(X) = 0$, the system is *DOWN*.
Definition. $P(\phi(\mathbf{X}) = 1) = R$ is called system reliability.

14. Prove that $R = E[\phi(\mathbf{X})]$. Using Problem 6, find the reliability of series and parallel systems.

15. A system has two minimal path sets $(1, 2, 3)$ and $(2, 4)$. Write system structure function and find system's reliability.

16. *Ushakov-Litvak bounds on system reliability* [27,28]. Let $L_1, L_2, ..., L_s$ be a collection of *non intersecting* minimal path sets of the system. Let $C_1, C_2, ..., C_k$ be a collection of *non intersecting* minimal cut sets of the system. (It is assumed that the system has independent binary components.) Prove the following Ushakov-Litvak bounds:

$$LB = 1 - \prod_{j=1}^{s}(1 - \prod_{i \in L_j} p_i) \le R \le \prod_{j=1}^{k}[1 - \prod_{i \in C_j}(1 - p_i)] = UB. \tag{2.5.5}$$

Remark. Since there might exist *several* collections of non intersecting paths and non intersecting cuts, it is possible to construct several lower bounds and use the *largest* of them. Similarly, one can construct several upper bounds and use the *smallest* of them. This method is in fact the Ushakov-Litvak original suggestion.#

17. Let $\mathbf{x} = (x_1, x_2, ..., x_n)$ and $\mathbf{y} = (y_1, y_2, ..., y_n)$ be binary vectors and let $\phi(\cdot)$ be system binary structure function. Let $\mathbf{z} = (z_1, z_2, ..., z_n)$, where $z_i = 1 - (1 - x_i)(1 - y_i)$, $i = 1, 2, ..., n$. Prove that $\phi(\mathbf{z}) \ge 1 - (1 - \phi(\mathbf{x}))(1 - \phi(\mathbf{y}))$.

18. For a given monotone system there are two options:
(i) Each element is duplicated, i.e. an identical element is connected in parallel to each system element.
(ii) The whole system is duplicated by another and similar one, and these two systems are connected in parallel.
Which option is preferable from the reliability point of view ?

Hint. Use the result of the previous problem.

19. Suppose we have a monotone binary system of n components. The structure function of this system is $\phi(x) = h(x_1, x_2, ..., x_n)$. Suppose that component i is replaced by another binary monotone system with structure function $\phi_i(y^{(i)})$, where $y^{(i)}$ is a binary vector, $i = 1, 2, ..., n$. For example, the original system is a parallel connection of two components. Suppose we decide to replace each component by a bridge structure. As a result, we will have a parallel connection of two bridge systems.

Find the general expression for the structure function after the above described replacement.

Remark. A particular case of the above described situation is a replacement of each component of the system by a subsystem identical to the original system. Such systems are called *recurrent*, see [27], Section 9.3.

20. Checking network connectivity. You are given a network with five nodes numbered $1, 2, 3, 4, 5$, and 6 edges. Each edge is represented by a five-dimensional 0/1 vector. For example, edge e_1 connecting nodes 1 and 2 is the following vector: $\{1, 1, 0, 0, 0\}$. There are five additional edges described as $e_2 = \{1, 0, 1, 0, 0\}$, $e_3 = \{0, 0, 0, 1, 1\}$, $e_4 = \{0, 0, 1, 1, 0\}$, $e_5 = \{0, 1, 0, 0, 1\}$, $e_6 = \{0, 1, 0, 1, 0\}$.

Write a *Mathematica* program for checking the network connectivity.

Hint. If two edge vectors have a node in common, their scalar product is positive. If they have no nodes in common, their scalar product is zero. For example, $\{1, 1, 0, 0, 0\} . \{1, 0, 1, 0, 0\} = 1$. (Dot means scalar product.) Indeed, both edges are incident to node 1. Contrary to that, $\{1, 1, 0, 0, 0\} . \{0, 0, 0, 1, 1\} = 0$.

Let us introduce a "gluing" operation of two vectors. If the scalar product of two vectors is not zero, these two vectors are glued together just by summation. So, gluing together edge e_1 and e_2 gives the vector $\{2, 1, 1, 0, 0\}$. The nonzero elements of this vector describe a connected component of edges e_1 and e_2.

If the scalar product of e_i and e_j is zero, "gluing" e_j to e_i leaves e_i *unchanged*.

The program must implement the following algorithm for testing connectivity. Fix vector e_1. Glue to it all other vectors $e_2, e_3, ..., e_6$. Then fix vector e_2. Glue to it all other vectors, etc. Finally, fix e_6 and glue to it all other vectors.

If after these operations there will be at least one vector with *all positive* components, then the network is connected. If all vectors have at least one zero component, the network is disconnected. Check this fact by considering several examples.

21. MC program for estimating probability of network connectivity. For the network described in the previous problem, you are given the probabilities p_i that edge i is *up*, $i = 1, ..., 6$. Edges are assumed to be independent. Complement the program which you have designed to solve the previous problem, to estimate the probability that the network is connected.

22. CMC for estimating network reliability with unreliable edges. To estimate network reliability by CMC you have to realize the following computational scheme.

1. For each edge, simulate its state (assuming the *up* probability is given).
2. Using DSS check the state of the network.
3. If the network state is UP add 1 to the appropriate variable.
4. Repeat 1-3 M times.
5. Compute the reliability estimate.

Develop a pseudocode for the above scheme.

Chapter 3

Exponentially Distributed Lifetime

3.1 Characteristic Property of the Exponential Distribution

The notation $\tau \sim Exp(\lambda)$ means that

$$F_\tau(t) = P(\tau \le t) = 1 - e^{-\lambda t}, t > 0. \tag{3.1.1}$$

The density function of r.v. τ is $f_\tau(t) = \lambda e^{-\lambda t}$.

The characteristic property of the exponential distribution is that the so-called *failure rate* $h(t)$ is **constant**:

$$h(t) = \frac{f_\tau(t)}{1 - F_\tau(t)} = \lambda = Const. \tag{3.1.2}$$

The probabilistic meaning of (3.1.2) is the following. Suppose we know that a component whose lifetime $\tau \sim Exp(\lambda)$ survived time t, i.e. $\tau > t$. Then the conditional probability to fail in the interval $[t, t + \delta t]$ does not depend on t:

$$P(\tau \in [t, t + \delta t] | \tau > t) = \frac{e^{-\lambda t} - e^{-\lambda(t+\delta t)}}{e^{-\lambda t}} = 1 - e^{-\lambda \delta t} \simeq \lambda \cdot \delta t.$$

Very often, it is said that an object whose lifetime has Exponential distribution is not aging: if it is "alive" at time t, its chances to survive past $t + T$ are exactly the same as the chances of a "newborn" object to remain alive during $[0, T]$.

We advise the reader to comprehend this amazing property which is characteristic only for an Exponential distribution. It is worth mentioning that this distribution is a continuous version of Geometric distribution which you definitely have studied in the Probability course.

In future, we will often use the probability that an object which is alive at time t will survive another "small" time interval δt. Formally,

$$P(\tau > t + \delta t | \tau > t) = \frac{P(\tau > t + \delta t)}{P(\tau > t)} = \exp[-\lambda \cdot \delta t] = \quad (3.1.3)$$
$$= P(\tau > \delta t) = 1 - \lambda \delta t + o(\delta t), \text{ as } \delta t \to 0.$$

Naturally, the complementary probability to fail in the interval $[t, t + \delta t]$ given the object is alive at t is $\lambda \delta t + o(\delta t)$.

3.2   Random Jump Process with Exponentially Distributed Sojourn Times

Let $[\xi(t), t \geq 0]$ be a continuous random process with discrete set of states $\{0, 1, 2, ...\}$. Assume that $\xi(t)$ sits in state i random time $\tau_i \sim Exp(\lambda_i)$, and afterwards moves into state $i + 1$. Figure 3.1 shows a trajectory of $\xi(t)$. We will assume that the r.v.'s τ_i are independent.

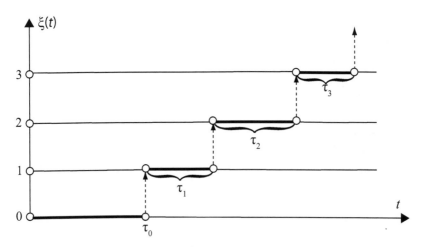

Figure 3.1: Trajectory of a monotone jump process

We will be interested in finding the probability that at time θ the process is in state k.

Claim 3.2.1

$$P(\xi(\theta) = k) = P(\xi(\theta) \geq k) - P(\xi(\theta) \geq k + 1). \qquad (3.2.1)$$

The proof is very simple. The first term in the right-hand side of (3.2.1) is the probability to be at time θ in one of the states $k, k+1, k+2, ...$ The second term is the probability to be at time θ in one of the states $k+1, k+2, ...$ Thus the difference of these terms is the probability to be in state k at time θ.#

It is important to note that Claim 1 remains true for *arbitrary distributions* of the "sitting" times τ_i.

Next important notion is the *convolution* of several c.d.f.'s. Suppose that $\tau_i \sim F_i(t), i = 1, ...k.$

Definition 3.2.1

The CDF of the sum $\sum_{i=1}^{k} \tau_i$, where τ_i are i.r.v.'s is called the *convolution* of $F_1(t), ..., F_k(t)$ and is denoted as $F^{(k)}(t).$#

The analytic expressions for the convolutions are as follows.

$$F(t) \equiv F^{(1)}(t) = \int_0^t f_1(x)dx, \quad F^{(2)}(t) = \int_0^t F^{(1)}(t-x)f_2(x)dx, \quad (3.2.2)$$

$$F^{(k)}(t) = \int_0^t F^{(k-1)}(t-x)f_k(x)dx.$$

It is easy to interpret these formulas using probabilistic arguments. Try it as an exercise!

For the case of identically distributed exponential i.r.v.'s it is possible to find an explicit analytic expression for the convolution. So, let $\tau_i \sim Exp(\lambda)$. Then

$$P(\tau_1 + ... + \tau_k \leq t) = 1 - e^{-\lambda t} \sum_{i=0}^{k-1} \frac{(\lambda t)^i}{i!}. \qquad (3.2.3)$$

This is so-called Gamma distribution with parameters (k, λ). It will be denoted as $Gamma(\lambda, k)$.

Let us consider now a connection between convolutions and the number of jumps in a random process. Suppose that our process sits in state 1 time τ_1, then jumps into state 2 and sits there random time τ_2, etc. R.v.'s τ_i may be *arbitrary* distributed and not necessarily *independent*. Denote by $N(t)$ the number of jumps of our random process on the interval $(0, t]$. The probabilistic meaning of the convolution is clarified by the following

Claim 3.2.2

$$P(\tau_1 + ... + \tau_k \leq t) = P(N(t) \geq k). \tag{3.2.4}$$

The proof is obvious: if the sum of k sitting times is $\leq t$, then there are at least k jumps on $(0, t]$. If, on the other hand, $N(t) \geq k$, the k-th jump took place before t or at t, which implies that $\tau_1 + ... + \tau_k \leq t.\#$

Remark. Let us return to the process $\xi(t)$, $t \geq 0$. Suppose the sitting time in state $\xi(t) = 0$ equals $\tau_0 = t_0$. Where is the process at time t_0? This is the matter of agreement. We will assume that the trajectories of $\xi(t)$ are *left-continuous*. In other words, if t_0 is the jumping time, at time t_0 the process state is defined to be 0. Similarly, if the jump from any state k into $k + 1$ appears at time $t = t^*$, we say that $\xi(t^*) = k$. From numerical point of view, our assumption *has no significance* because the sitting times are continuous random variables.#

Let us prove the following useful
Lemma 3.2.1

$$P(\xi(t_0) = k) = P(\tau_0 + ... + \tau_k \leq t_0) - P(\tau_0 + ... + \tau_k + \tau_{k+1} \leq t).(3.2.5)$$

The proof is left as an exercise.

3.3   Examples

In this section we consider several simple examples of random processes involving exponential distribution.

Example 3.3.1. Two-component system.

A system consists of two independent components. Component i has lifetime $\tau_i \sim Exp(\lambda_i), i = 1, 2$. At $t = 0$ both components are operating, i.e. they are *up*. Denote this state of the system by (1,1). At some random moment, one of the components breaks down, and the system moves into one of the states (0,1) or (1,0). (State (0,1) means that the first component is *down*, and the second is *up*.) After some random time, the system moves from any of these two states into the state (0,0) - the state of system failure, see Fig. 3.2. (We can describe this system in terms of a jump process $\xi(t)$ with four states: $(1, 1), (0, 1), (1, 0), (0, 0)$ corresponding to $\xi = 0, 1, 2, 3$, respectively. Here the transitions $0 \to 1$, $0 \to 2$, $1 \to 3$, $2 \to 3$ take places.)

Questions: how is the sitting time distributed in state (1,1)? What is the probability that the system will move from state (1,1) into (0,1) or into

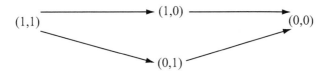

Figure 3.2: Transition diagram for Example 3.3.1

(1,0)? Will this probability depend on the sitting time in (1,1)? How is the sitting time distributed in state $(0,1)$ or $(1,0)$?

Solution
Obviously, the system stays in state (1,1) random time $\tau(1,1) = \min(\tau_1, \tau_2)$. Note that

$$P(\tau(1,1) > t) = P(\tau_1 > t, \tau_2 > t) = [\text{because of independence}] \quad (3.3.1)$$
$$= P(\tau_1 > t)P(\tau_2 > t) = e^{-t(\lambda_1 + \lambda_2)}.$$

It follows, therefore that $\tau(1,1) \sim Exp(\lambda_1 + \lambda_2)$.
  Now the system moves from (1,1) to (0,1) iff $\tau_1 < \tau_2$.

$$P(\tau_1 < \tau_2) = \int_0^\infty f_1(t)(1 - F_2(t))dt = \quad (3.3.2)$$
$$\int_0^\infty \lambda_1 e^{-\lambda_1 t} e^{-\lambda_2 t} dt = \frac{\lambda_1}{\lambda_1 + \lambda_2}.$$

The above example is quite elementary but very useful. As an exercise, try to generalize it to a system with k components. Prove that the sitting time in any state with k "alive" components is exponentially distributed with parameter $\Lambda = \sum_{i=1}^{k} \lambda_i$, where λ_i-s are the failure rates of the components alive.

  As an exercise prove also that the probability of transition from state (1,1,...,1) into state (0,1,...,1) is equal to λ_1/Λ.

  Hint. $P(\tau_1 < \tau_i \text{ for all } i > 1) = P(\tau_1 < \min\{\tau_i\}, i = 2, ...k)$.

Example 3.3.2. Two-state alternating process.

  A system has two states "*UP*" and "*DOWN*". At $t = 0$ the system is in the *UP* state. It remains in this state (operation state) random time $\tau_u \sim Exp(\lambda)$ and then moves into state *DOWN* (state of repair). The sitting time in *DOWN* is also exponentially distributed $\tau_d \sim Exp(\mu)$. System behavior is illustrated in Fig. 3.3.

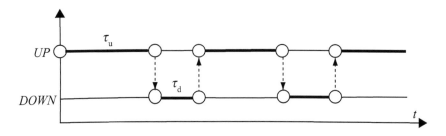

Figure 3.3: Trajectory of a two-state alternating process

Our goal is to find $P_u(t) = P(\text{the system is } UP \text{ at time } t)$. Obviously, $P_d(t) = P(\text{the system is } DOWN \text{ at time t}) = 1 - P_u(t)$. Besides, we want to know the so-called *stationary probability* that the system is UP at some remote time t, $t \to \infty$.

The derivation is very instructive because it demonstrates the probability balance equation technique and relies heavily on the fact that both *UP* and *DOWN* periods are exponentially distributed. Experienced reader may skip this material.

$$P_u(t + \delta t) = P_u(t)P(\text{stay in } UP \text{ during}(t, t + \delta t)|\ UP \text{ at } t) + (3.3.3)$$
$$P_d(t)P(\text{move from } DOWN \text{ to } UP \text{ in } (t, t + \delta t)|\ DOWN \text{ at } t) =$$
$$P_u(t)(1 - \lambda\delta t) + P_d(t)\mu\delta t + o(\delta t).$$

After simple algebra, letting $\delta t \to 0$, it is easy to derive the following differential equation:

$$\frac{dP_u(t)}{dt} = -P_u(t)\lambda + P_d(t)\mu. \tag{3.3.4}$$

This equation must be solved under conditions $P_u(0) = 1, P_d(0) = 0, P_u(t) + P_d(t) = 1$.

We present the final result:

$$P_u(t) = \frac{\mu}{\lambda + \mu} + \frac{\lambda}{\lambda + \mu}e^{-(\lambda+\mu)t}, \tag{3.3.5}$$
$$P_d(t) = \frac{\lambda}{\lambda + \mu} - \frac{\lambda}{\lambda + \mu}e^{-(\lambda+\mu)t},$$

We advise the reader to analyze the relationship (3.3.3). The key is to understand why P(move from $DOWN$ to UP in $(t, t+\delta t)$|to be in $DOWN$ at t) $= \mu\delta t + o(\delta t)$, and to understand that this is true because $\tau_d \sim Exp(\mu)$.

Note that $P_u(t)$ tends to $\mu/(\lambda + \mu)$ as t tends to infinity. This is the so-called *stationary probability* to be in the UP state. Also important is to note that this probability (called often *availability* and denoted Av) can be expressed via the mean values of the sitting times, as follows:

$$Av = \frac{\text{mean time in } UP}{\text{mean time in } UP + \text{mean time in } DOWN}. \qquad (3.3.6)$$

(We remind that $E[\tau_{up}] = 1/\lambda$ and that $E[\tau_d] = 1/\mu$.)

Remark 1. The formula (3.3.6) is true for arbitrary distributed UP and $DOWN$ times. To prove this, we need some facts from Renewal Theory, see e.g. [2,20]. We will omit the formal proof.#

Example 3.3.3. Poisson process
There are several equivalent ways to define the Poisson process. One of the simplest ways is the following
Definition 3.3.1. The jump process $[\xi(t),\ t > 0]$ with the state space $\{1, 2, 3, ...\}$ and independent sitting times in each state i, $\tau_i \sim Exp(\lambda)$ is called a Poisson process with parameter λ.#

Poisson process has the following interesting properties:
(i) The probability that a jump will appear in the small interval $[t, t+\Delta]$ is $\lambda\Delta + o(\Delta)$; it does not depend on when and how many jumps were before t.

(ii) The number of jumps in the interval $[0, T]$ has a Poisson distribution with parameter $\Lambda = \lambda T$, i.e. $P(N(T) = k) = \exp[-\Lambda]\Lambda^k/k!,\ k = 0, 1, 2, ...$

(iii) The mean number of jumps on $[0, T]$ equals $E[N(T)] = \Lambda = \lambda T$.

More information on Poisson process can be found in many textbooks, see for example [20].

Remark 2. *Markov property, Markov process.* The jump process $[\xi(t),\ t > 0]$ considered in the definition of Poisson process has the following fundamental property. If it is known that $\xi(\theta) = m$, then the probability of any event in the future of $\xi(t)$, i.e. in the interval $[\theta, \infty]$ *does not depend* on the trajectory of $\xi(t)$ before θ, i.e. on the history on $[0, \theta]$. This is the Markov property and the Poisson process possesses it. Loosely speaking, we say that Markov property means that the future depends only on the actual state and not on the previous history. The two-state process considered in Example 3.3.2 is also a Markov process.

It is easy to give an example of a jump process which is not a Markov process. Suppose that jumps appear after regular intervals of length a. If we know that the process is now in a certain state, the time of the nearest jump *does depend* on the past, i.e. on the time when the previous jump took place. The Markov property of Poisson process is provided by the amazing memoryless properties of the exponential distribution.#

3.4 Problems and Exercises

1. A system consists of three components. Component i has failure rate λ_i, $i = 1, 2, 3$. At $t = 0$ all components are *up*. Find the expression for the probability that the components fail in the following order : $3 \rightarrow 2 \rightarrow 1$.

2. Let $X \sim U(0, 1)$. Find the expression for the failure rate.

3. Component has failure rate $h(x)$ and the c.d.f. $F(t)$. Prove the following formula:

$$1 - F(t) = e^{-\int_0^t h(x)dx}. \tag{3.4.1}$$

4. Let $X \sim Gamma(n, \lambda)$ and $Y \sim Gamma(m, \lambda)$. How is $Z = X + Y$ distributed? (X and Y are independent).

5. Let $X \sim U(0, 1)$. How is $Y = -\log(X)$ distributed?

6. Derive (3.2.2) by using probabilistic arguments.

7. Suppose that the system has n independent components, the i-th component has lifetime $\tau_i \sim Exp(\lambda_i)$. At $t = 0$ all components are *up*. Find the probability that component 1 will fail while all other components are up.

8. *Geometric distribution.* Let X be a discrete random variable taking on values $k \geq 1$. $P(X = k) = q^{k-1}p$, $q = 1 - p$. We say that $X \sim G(p)$, i.e. X is geometrically distributed.

This distribution is a discrete analogue of exponential distribution.

Prove that $P(X > M) = q^M$ and derive from here the discrete analogue of memoryless property:

$$P(X > M + N | X > M) = P(X > N).$$ What is the physical model behind this distribution?

9. *Two-component series system.* Suppose we have a system of two independent components whose failure takes place if any of them fails. One can think of a series system as a three-node network with reliable nodes

s, a, t and unreliable edges $e_1 = (s, a), e_2 = (a, t)$. The network fails if the connection $s - t$ is broken. $\tau(e_1) \sim F_1(t), \tau(e_2) \sim F_2(t)$.

9.1. Prove that the lifetime of the whole system has the c.d.f. $F(t) = 1 - (1 - F_1(t))(1 - F_2(t))$.

9.2. Assuming that $F_i(t)$ has density $f_i(t)$, derive an expression for the failure rate of the system and prove that it equals to the sum of failure rates of the components.

10. $\tau_i \sim Exp(\lambda), i = 1, 2, 3$ and are independent random variables. Prove that $P(\tau_1 \le \tau_2 \le \tau_3) = 1/6$.

11. Suppose a system consists of k exponential components, component i has failure rate λ_i, $i = 1, 2, ..., k$. At time $t = 0$ all components are *up*. Denote by τ the random time of the appearance of the first failure. Prove that $\tau \sim Exp(\sum_{i=1}^{k} \lambda_i)$. Prove that the probability that component j fails first equals $\lambda_j / \sum_{i=1}^{k} \lambda_i$.

12. Suppose that in Problem 11 all $\lambda_i = \lambda$. Find out the mean time to the failure of all k components of the system.

Hint. Use the following very convenient formula for computing the mean of nonnegative r.v. X with CDF $F(t)$:

$$E[X] = \int_0^\infty (1 - F(x))dx. \tag{3.4.2}$$

13. Using the model of Gamma distribution prove that if $X \sim Gamma(n, \lambda)$, then $E[X] = n/\lambda$.

14. Weibull distribution. Suppose that a nonnegative random variable X has failure rate $h_X(t) = \alpha t^{\beta-1}$. Derive the CDF of X and the corresponding density function. Weibull distribution family is very popular in reliability theory since for $\beta > 1$ it gives an increasing failure rate, so-called *aging*, for $\beta < 1$, the failure rate is decreasing, and for $\beta = 1$ it becomes exponential distribution.

15. Find the CDF of a system which is a series connection of two components having failure rates $h_i(t) = \alpha_i t^{(\beta-1)}$ $i = 1, 2$

16. Let $X \sim U(0, 1)$. Describe a method of generating, using X, Weibull random variable with parameters (α, β).

Hint. Use the following fact: solving the equation (with respect to Y)

$$F(Y) = X, \tag{3.4.3}$$

where $F(\cdot)$ is some continuous CDF, produces random variable $Y \sim F(\cdot)$.

17. Prove the Lemma in Section 3.2.

Chapter 4

System Static and Dynamic Reliability

Everything should be made as simple as possible, but not simpler.

Albert Einstein

4.1 System Description. Static Reliability

The system consists of n components (elements), each component has two states : *up* (denoted by 1) and *down* (denoted by 0). The state of the whole system is described by a vector $e = (e_1, e_2, ..., e_n)$, where e_i is 0 or 1 for component i being *down* or *up*, respectively.

The system can be in two states too, *UP* and *DOWN*, denoted also by 1 or 0, respectively. System state is expressed via binary function $\phi(e)$. The set of all those vectors e for which the system is *UP* is denoted by Up:

$$Up = \{e : \phi(e) = 1\}. \tag{4.1.1}$$

$\phi(e)$ is called system *structure function*.

The set of all n-digit binary vectors is denoted by Ω. The complement of Up will be denoted as *Down*:

$$Down = \{e : \phi(e) = 0\}. \tag{4.1.2}$$

Obviously $Down \bigcup Up = \Omega$.

We consider only *monotone* systems, which means the following.

$$(0, 0, ...0) \in Down, \ (1, 1, ...1) \in Up, \ v \leq w \rightarrow \phi(v) \leq \phi(w). \tag{4.1.3}$$

In words: the system state cannot get worse if a component changes its state from down to up.

Finally, for each state vector e we denote by $U(e)$ and $D(e)$ the sets of indices corresponding to the *up*-components of e and *down*-components of e, respectively. For example, if $e = (1,1,0,0,1,0,1,0)$ then $U(e) = \{1,2,5,7\}, D(e) = \{3,4,6,8\}$

So far we did not introduce probability. Suppose now that component i is *up* with probability p_i and *down* with complementary probability $q_i = 1 - p_i$. All components are assumed to be *statistically independent*. Then the probability that the system is in state $e = (e_1, ..., e_n)$ equals to

$$P(e) = \prod_{i \in U(e)} p_i \prod_{i \in D(e)} q_i. \qquad (4.1.4)$$

Finally, the probability R_s that the system is UP is determined as

$$R_s = \sum_{e \in Up} P(e) = \sum_{e \in Up} \left(\prod_{i \in U(e)} p_i \prod_{i \in D(e)} q_i \right). \qquad (4.1.5)$$

This formula looks complex. To see that it is in fact quite simple, let us consider an example. Suppose we have a system of four components. It is defined to be UP if and only if at least three of the components are up. These states are $(1,1,1,0), (1,1,0,1), (1,0,1,1), (0,1,1,1)$, and $(1,1,1,1)$. Suppose p is the up probability for all components, $q = 1 - p$. Then

$$R_s = 4p^3(1-p) + p^4. \qquad (4.1.6)$$

Remark. The reader who solved all or almost all problems in Chapter 2 will probably recognize that the notions of monotone binary systems and structure function are already familiar to her/to him.#

We will go a little further in describing the binary systems with binary components. Introduce a binary variable X_i for component i, put $P(X_i = 1) = p_i$ and $P(X_i = 0) = 1 - p_i = q_i$. Further, we define system state as a binary random variable $\phi(\mathbf{X})$ (called *structure function*), where $\mathbf{X} = (X_1, X_2, ..., X_n)$. The dependence of system state on the component states is described via some function $\Psi(\cdot)$:

$$\phi(\mathbf{X}) = \Psi(X_1, X_2, ..., X_n). \qquad (4.1.7)$$

In the problems of Chapter 2 we presented several examples of $\Psi(\mathbf{X})$ based on minimal paths and minimal cuts.

$\Psi(\mathbf{X})$ is a binary random variable. It leads to the following definition:

Definition 4.1.1

$$P(\Psi(\mathbf{X}) = 1)) = P(system\ is\ UP) = R_s \qquad (4.1.8)$$

is called system *static reliability*.

Since $\Psi(\mathbf{X})$ is a binary random variable, the static reliability can be expressed via the operation of finding its mean value:

$$R_s = E[\Psi(\mathbf{X})] = E[\Psi(X_1, X_2, ..., X_n)]. \qquad (4.1.9)$$

This formula is very useful and efficient if we can derive an analytic expression of the structure function. Unfortunately, this expression is available only for a rather simple system with small number of components.

4.2 Dynamic Reliability

Now let us involve *time* and investigate how system reliability depends on time. For that purpose we first introduce random component lives. Suppose that component i is up at time $t = 0$, and remains in this state for a random time $\tau_i \sim F_i(t)$. After random time τ_i the component gets down and remains down "forever". Denote

$$R_i(t_0) = P(\tau_i > t_0) = 1 - F_i(t_0); \quad Q_i(t_0) = F_i(t_0). \qquad (4.2.1)$$

Obviously, $R_i(t_0)$ is the probability that component i is up at time t_0 and therefore was up during the whole interval $[0, t_0]$. Then, obviously, the probability that the system is UP at the instant t_0 is equal to

$$R_s(t_0) = \sum_{e \in Up} \left(\prod_{i \in U(e)} R_i(t_0) \prod_{i \in D(e)} Q_i(t_0) \right). \qquad (4.2.2)$$

Note the following important fact: If we have a monotone system, and it is UP at time instant t_0, then it was UP during the whole period $[0, t_0]$ and therefore, $R_s(t_0)$ is the probability that system random lifetime τ_s is greater than t_0:

$$R_s(t_0) = P(\tau_s > t_0), \ P(\tau_s \le t_0) = F_s(t_0) = 1 - R_s(t_0). \qquad (4.2.3)$$

If we compare formulas for R_s and $R_s(t_0)$ we can formulate the following rule.

For a monotone system, the reliability as a function of time can be obtained from the expression of system static reliability by replacing p_i by $R_i(t)$ and q_i by $Q_i(t)$.

Remark 1. Let $E[X_i] = p_i$. It is tempting to write $E[\Psi(\mathbf{X})] = \Psi(p_1, p_2, ..., p_n)$. In general, this is *not* true. For example, let $\Psi(\mathbf{X}) = 1 - (1 - X_1 X_2)(1 - X_1 X_3) = X_1 X_2 + X_1 X_3 - X_1 X_2 X_3$. Then $E[\Psi(\mathbf{X})] = p_1 p_2 + p_1 p_3 - p_1 p_2 p_3 \neq 1 - (1 - p_1 p_2)(1 - p_1 p_3)$! To avoid possible errors we will write $R_s = E[\Psi(X)] = \Psi_0(p_1, p_2, ..., p_n).\#$

Suppose we have an explicit expression for $R_s(t) = P(\tau_s > t)$. Then there is an easy way to find the mean system UP time, i.e. the mean value of the interval during which the system is in UP state:

$$E[\tau_s] = \int_0^\infty R_s(t)dt. \tag{4.2.4}$$

This follows from the fact known from Probability Theory: If τ is a *nonnegative* random variable with CDF $F(t)$ then $E[\tau] = \int_0^\infty (1 - F(t))dt$. We omit the proof of this useful formula.

Example 4.2.1

Consider a series system of n independent components. $\tau_i \sim Exp(\lambda_i)$. As an exercise, derive the expression for the CDF of system lifetime.

Hint. By definition, a series system is UP if and only if all its components are *up*.$\#$

Remark 2. For a series system, its lifetime is a <u>minimum</u> of independent r.v.'s $\tau_1, ..., \tau_n.\#$

Example 4.2.2

Derive a formula for the CDF of system lifetime for a parallel system. A parallel system, by definition, is $DOWN$ if all its components are *down*.

Remark 3. Note that in Example 4.2.2 system lifetime is <u>maximum</u> of r.v.'s $\tau_1, ..., \tau_n.\#$

4.3 Stationary Availability: Systems With Independent Renewable Components

Now we will consider a system of n independent components, each of which operates in the following way: component i works time $\tau_i \sim F_i(t)$, then undergoes repair during random time $\omega_i \sim G_i(t)$, which is independent on the operation time, then again operates, is repaired, etc. We have already considered such a system in Chapter 3, Example 3.3.2. If the operation and

repair times are exponential, it has been proved that the probability that
the component is *up* at time $t, t \to \infty$, is

$$Av_i = \frac{E[\tau_i]}{E[\tau_i] + E[\omega_i]}. \qquad (4.3.1)$$

Note that (4.3.1) holds true for *arbitrary* distributions of operation and
repair times. This result follows from renewal theory and we will not go
here into the details of the proof.

Now let us return to the two-state alternating process considered in
Chapter 3. Denote by $Av_i(t)$ the probability that the component i is *up* at
the instant t. Now take the expression (4.1.5) for system static probability
and replace p_i by $Av_i(t)$ and q_i by $1 - Av_i(t)$. Then the formula (4.1.5)
will express the "instant" probability that the system is UP at time t. Now
if we let $t \to \infty$, each $Av_i(t) \to Av_i$, and R_s will tend to a limit which
expresses system stationary availability Av_s. Av_s has two interpretations:
it equals the probability that the system is UP at some remote time instant
t; it equals also to the ratio of the mean stationary UP time to the sum of
system mean stationary UP time and mean stationary $DOWN$ time.

4.4 Burtin-Pittel Approximation to Network Reliability

To explain the main idea of the Burtin-Pittel (B-P) approximation, assume
that all components have failure probability β. Consider any failure state
e of the system. The probability that the system is in this state equals
$Q(e) = \beta^k(1 - \beta)^{n-k}$, where k is the number of down components in the
state e: $k = |D(e)|$. After simple algebra we obtain that for $\beta \to 0$,
$\quad Q(e) = \beta^k + O(\beta^{k+1}) = \beta^k + o(\beta^k)$.
Since β is "small", the main term in $Q(e)$ is β^k. Now among all failure
states, the <u>largest</u> failure probability will have the states for which $|Q(e)|$ is
the <u>smallest</u>. The failure probabilities of such states make the main contribution to system failure probability.

Example 4.3.1

Let us return to example of a four-component system which fails if two
or more of its components fail. There are 6 failure states with exactly 2
failed components, 4 failure states with exactly 3 failed components and 1
failure state with 4 failed components. (Check it!) Thus the probability
that the system is $DOWN$ equals

$P(DOWN) = 6q^2(1-q)^2 + 4q^3(1-q) + q^4.$

After simple algebra we obtain the following expression

$$P(DOWN) = 6q^2 - 8q^3 + 3q^4. \tag{4.4.1}$$

Now the key point is that for $q \to 0$, the main term in the system down probability is $6q^2$ and all other terms are of smaller magnitude. Formally,

$$P(DOWN) = 6q^2 + o(q^2). \tag{4.4.2}$$

The physical meaning of this result will become clear if we consider the network which corresponds to the system under consideration. On Fig. 4.1 we see a 4-node and 4-edge network.

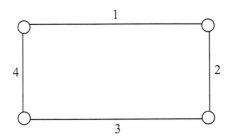

Figure 4.1: Network becomes disconnected if any pair of edges fail

Network failure is defined as loss of connectivity. There are exactly 6 pairs of edges whose failure leads to network failure :

$(1,2), (1,3), (1,4), (2,3), (2,4), (3,4),$

and 6 is the coefficient at q^2 in the expression (4.4.1)!#

The Burtin-Pittel approximation is in fact a generalization of the above approach.

Now let us proceed more formally and consider the following situation:

(i) System components are nonrenewable and have exponential lifetimes $\tau_i \sim Exp(\lambda_i)$.

(ii) Components are highly reliable, i.e. the λ_i are small.

This is formalized as follows. We assume that

$$\lambda_i = \alpha \cdot \theta_i, \quad \alpha \to 0. \tag{4.4.3}$$

An immediate consequence of (ii) is that for fixed t, as $\alpha \to 0$,

$$P(\tau_i > t) = e^{-\lambda_i t} = 1 - \alpha\theta_i t + O(\alpha^2), \quad P(\tau_i \le t) = \alpha\theta_i t + O(\alpha^2). \tag{4.4.4}$$

Let us now consider the expression for the probability that the system is down at fixed instant t, $1 - R_s(t)$, and use the condition (ii). To get the desirable analytic form, let us partition the whole set of system DOWN states $D = \Omega - Up$ into subsets $D_r, D_{r+1}, ...$ according to the number of the failed components k in the system state vectors, $k = r, r+1, ...n$. Obviously,

$$D = \bigcup_{k=r}^{n} D_k.$$

Note that D_r is the set of all down states with the smallest number of failed elements r.

$$1 - R(t) = \sum_{k=r}^{n} \left(\sum_{e \in D_k} \left(\prod_{i \in U(e)} e^{-\lambda_i t} \cdot \prod_{i \in D(e)} (1 - e^{-\lambda_i t}) \right) \right). \tag{4.4.5}$$

Indeed, the double sum in (4.4.5) is the probability that the system is in one of its down states. For each such state e, the elements of $U(e)$ are up and the elements of $D(e)$ are down. Now use (ii).

The main term in (4.4.5) will be determined by the first summand of the internal sum with the smallest $k = r$. After some algebra it follows that for $\alpha \to 0$

$$1 - R(t) = \alpha^r \cdot t^r \cdot g(\theta) + O(\alpha^{r+1}) \approx 1 - \exp(-\alpha^r t^r g(\theta)), \tag{4.4.6}$$

where $g(\cdot)$ is the sum of the products of θ_i over all failure states with minimal number r of failed components:

$$g(\theta) = \sum_{e \in D_r} \prod_{i \in D(e)} \theta_i. \tag{4.4.7}$$

Let us consider an example illustrating the use and the quality of the approximation (4.4.6).

Example 4.4.1: $s - t$ **connectivity of a dodecahedron network.**
Fig. 4.2 shows a network with 20 nodes and 30 edges called *dodecahedron*. The nodes are absolutely reliable. The edges fail independently and their lifetimes $\tau \sim \text{Exp}(\lambda)$. The network fails if there is no path leading from node 1 ("source") to node 2 ("terminal"). The reliability criterion of such a network is termed the $s - t$ connectivity.

We assume that all $\theta_i = 1$ and α is "small". (Formally, $\alpha \to 0$.) The dodecahedron has two failure states with the smallest possible number of $r = 3$ edge failures. Indeed, node 1 is disconnected from node 2 if the edges $(1, 5, 26)$ fail, or the edges $(1, 2, 27)$ fail. All other failure states separating the source from the terminal have size greater than $r = 3$.

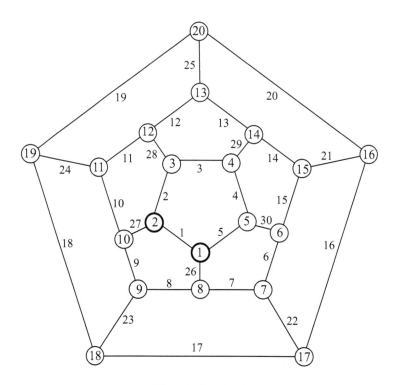

Figure 4.2: The dodecahedron network

Let us fix $t = 1$ and check how good is the approximation to network failure probability provided by the expression $F_{approx}(1) = 1 - \exp[\alpha^3 g(\theta)]$, for various values of α approaching zero. Since by (4.4.7) $g(\theta) = 2\theta^3 = 2$, our approximation is $F_{approx}(1) = 1 - \exp[-2\alpha^3]$.

Table 4.1 shows the values of network reliability by the B-P approximation versus the exact values of network failure probabilities $F_{exact}(1)$, for α ranging from 0.22314 to 0.01005. ($F_{exact}(1)$ was computed by J.S. Provan using an algorithm based on cut set enumeration, see [20].)

It is seen from Table 4.1 that for α below 0.05 the Burtin-Pittel (BP) approximation is quite satisfactory. More details and references on this approximation can be found in [19,20].#

Table 4.1: Comparison of exact and approximate reliability

$e^{-\alpha}$	α	$F_{approx}(1)$	$F_{exact}(1)$	rel. error in %
0.8	0.22314	0.0219	0.0362	39
0.85	0.16252	0.0854	0.0122	30
0.90	0.10536	0.00234	0.00288	18
0.92	0.08338	0.00116	0.00136	15
0.94	0.06188	0.000474	0.000528	10
0.96	0.04082	0.000136	0.000146	7
0.98	0.02020	$1.65 \cdot 10^{-5}$	$1.7 \cdot 10^{-5}$	3
0.99	0.01005	$2.0 \cdot 10^{-6}$	$2.03 \cdot 10^{-6}$	1.5

4.5 Pivotal Formula and Birnbaum Measure of Importance. Reliability Gradient Vector

We start with the representation of system static reliability R as a function of its component reliability p_i:

$$R = \Psi_0(p_1, p_2, ..., p_n). \tag{4.5.1}$$

Let us not worry that we may not know the exact analytic form of the Ψ function. All our conclusions we will get just from an assumption that such function does exist.

It is worth noting that in this formula R depends only on marginal probabilities p_i which is correct only for a system of *independent* components.

Now by the total probability formula,

$$R = P(\text{the system is } UP) = P(i \text{ is } up)P(\text{system } UP|i \text{ is } up) + \tag{4.5.2}$$
$$P(i \text{ is } down)P(\text{system } UP|i \text{ is } down).$$

This formula can be rewritten as

$$R = p_i \cdot \Psi_0(p_1, ..., p_{i-1}, 1, p_{i+1}, ..., p_n) + \tag{4.5.3}$$
$$q_i \cdot \Psi_0(p_1, ..., p_{i-1}, 0, p_{i+1}, ..., p_n).$$

Here $\Psi_0(p_1, ..., p_{i-1}, 1, p_{i+1}, ..., p_n)$ is system reliability in which the i-th component is replaced by absolutely reliable one. Similarly, $\Psi_0(p_1, ..., p_{i-1}, 0, p_{i+1}, ..., p_n)$ is the system reliability where the i-th component is permanently *down*.

These two operations are well understood for a system represented by a
network structure. If component i is an edge (a, b), and edges are subject
to failure, replacing (a, b) by absolutely reliable edge means compression of
nodes a and b into one. Replacing (a, b) by a permanently *down* edge means
elimination of (a, b) from the network. These two operations are illustrated
by Fig. 4.3.

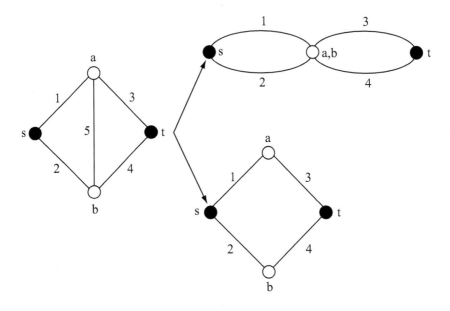

Figure 4.3: Pivotal decomposition of a bridge-type system. The pivoting is
around edge (a, b)

The formula (4.5.3) (called **pivoting** formula) is the core of one of the
first and rather powerful algorithms for computing $s - t$ networks reliability.
The process of pivoting continues until the algorithm leads to a series-parallel
subnetwork with no common components for which explicit formulas are
available for reliability calculations. In Fig. 4.3, pivoting around edge 5
leads immediately to series-parallel subsystems. Choosing the "best" edge
for pivoting is, however, algorithmically a quite difficult task.

To make easier the implementation of the pivoting algorithm, let us
remind the formulas for reliability of a series, parallel, and series-parallel
systems.

If our system consists of k subsystems connected in **series**, its reliability is a product of subsystems reliability (check it!):

$$R_{series} = \prod_{i=1}^{k} R_i, \tag{4.5.4}$$

where R_i is the reliability of i-th subsystem.

If a system is a **parallel** connection of k subsystems, its reliability equals (check it!)

$$R_{parallel} = 1 - \prod_{i=1}^{k}[1 - R_i]. \tag{4.5.5}$$

Definition 4.5.1. The expression

$$\frac{\partial \Psi_0(p_1, ..., p_n)}{\partial p_i} \tag{4.5.6}$$

is called **Birnbaum importance measure** (BIM) of component i.#

It follows from (4.5.3) that

$$\frac{\partial \Psi_0(p_1, ..., p_n)}{\partial p_i} = \Psi_0(p_1, ..., p_{i-1}, 1, p_{i+1}, ..., p_n) - \tag{4.5.7}$$
$$\Psi_0(p_1, ..., p_{i-1}, 0, p_{i+1}, ..., p_n).$$

Birnbaum measure of importance, suggested in [3], as it is seen from this expression, has a transparent physical meaning: it is the gain in system reliability received by replacing a *down* component i by an absolutely reliable one.

For the case of equal component reliability $p_i \equiv p$, first the partial derivatives (4.5.7) must be calculated and only afterwards all p_i are set to be equal p.

As an exercise consider a series and a parallel system of n components. Find out the Birnbaum measure of importance of component i and check that for series system, the most important is the less reliable component, and for parallel system the most important is the most reliable component, see Problem 10 in Section 4.6.

Definition 4.5.2. Reliability gradient vector ∇R is defined as

$$\nabla R = [\frac{\partial R}{\partial p_1}, ..., \frac{\partial R}{\partial p_n}].\# \tag{4.5.8}$$

In words: component i of the reliability gradient vector is component's i BIM.

The reliability gradient vector will be later used to synthesize network with optimal reliability parameters.

Remark 1. Several alternative importance measures (IM's) have been proposed in literature and used in reliability practice. One of them is called Fussel-Vesely Inportance Measure (FVIM) [18], and it is defined according to the following formula:

$$FV_i = 1 - \frac{R(p_1, p_2, ..., p_{i-1}, 0, p_{i+1}, ..., p_n)}{R(p_1, p_2, ..., p_n)}. \tag{4.5.9}$$

It quantifies the relative decrement in system reliability caused by a particular component failure. Contrary to the BIM and FVIM, which quantify the importance of individual components, several other IM's have been suggested to quantify the contribution to reliability of a *group* of components. They are known as Joint Importance Measures (JIM's), see [1]. The use of FVIM's and JIM's rests on the possibility to obtain in a closed form the expression for system reliability function $R(p_1, ..., p_n)$. For an important particular case of *equal* failure probabilities, the Birnbaum Importance Measure allows, via a specially designed combinatorial algorithm, to compute the component BIM's without having an analytic expression for the reliability function. We will discuss this topic in Chapter 10.

Remark 2. For systems with renewable components more suitable measure of component performance is so-called Barlow measure of performance. The gradient vector is also used in this measure. For more details see [20].#

Remark 3. Estimation of network reliability may be a quite difficult task, especially for large and highly reliable networks. The exact calculation needs special algorithms and software. Monte Carlo simulation is a good alternative, see e.g. Elperin et al. [11]. The dodecahedron example demonstrates that for very reliable networks with failure probability less than 10^{-4}, the Burtin-Pittel approximation provides with minimal efforts a reasonably accurate solution.#

4.6   Problems and Exercises

1. Let $\xi \sim U(0, 1)$. Prove that $X = -\log(1 - \xi) \sim Exp(1)$.

2. Let $\xi \sim U(0, 1)$. Prove that $X = -\lambda^{-1} \log(1 - \xi) \sim Exp(\lambda)$.

3. Let $X \sim F(t)$, $F(t)$ is a continuous increasing function. Define a new r.v. $Y = F(X)$. Try to prove Smirnov's theorem:
$$Y \sim U(0, 1).$$

4. Let $X \sim U(0, 1)$, $Y \sim U(1, 2)$. Denote by $F_X(\cdot)$ the c.d.f. of X, and by $F_Y(\cdot)$ the c.d.f. of Y. Let $Z \sim pF_X(t) + (1 - p)F_Y(t)$. Suggest an algorithm for simulating r.v. Z.

5. A small computer star-type network has 5 nodes: a, b, c, d, f. These nodes are connected by the following edges $(a, f), (b, f), (c, f), (d, f)$. Nodes are absolutely reliable, edges are subject to failures. Network fails if the central node f becomes disconnected from any of other nodes. Failure probabilities are 0.05 for first two edges and 0.03 for the last two. Each edge has a "spare" edge which can be put in parallel to this edge. These spare edges have failure probability 0.1. You are allowed to reinforce only one edge by a "spare" edge. What is your choice in order to achieve maximal reliability?

6. Solve Problem 5 for arbitrary edge up probabilities p_1, p_2, p_3, p_4. Find BIM for edge i.

7. You have a star-type network described in Problem 5. Each edge mean operation time equals 100 hours. After edge fails it undergoes repair which lasts on the average 5 hours. Find network instant availability.

8. The network has the following structure:

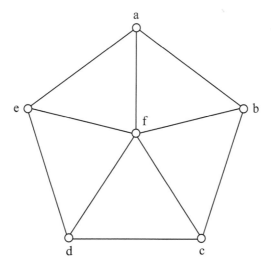

Figure 4.4: Pentagon-type star network

You have at your disposal five type-1 edges and five type-2 edges. Type-1 edges have failure probability 0.01; type-2 edges have failure probability 0.05. You are requested to design your network as to provide **minimal** network failure probability. (Network failure is defined as the loss of connectivity, i.e. separation of at least one node from other nodes.) To solve the problem you are allowed to use Burtin-Pittel formula for failure probability. What is the optimal location of edge types in the network?

9. If all $p_i = p$, is it true that BIM of component i is an increasing function of p?

Hint. Let us consider the system reliability function $R = \Psi_0(p_1, ..., p_n)$ (the components are independent). We already know that component i importance is defined as $\partial \Psi_0 / \partial p_i$. Let us add the condition that $p_i = p$ for all $i = 1, ..., n$.

Then component i importance will be the following function of p:

$$\frac{\partial \Psi_0}{\partial p_i}\Big|_{p_i=p, \, \forall i} = \text{Imp}_i(p). \tag{4.6.1}$$

We are interested in investigating the behavior of the function $\text{Imp}_i(p)$.

Let us show that this function is not necessarily monotone increasing function of p.

To do this we use the following property of the reliability function $\Psi_0(p) = \Psi_0(p_1, ...p_n)|_{p_i=p, \, \forall i}$, see [2], p.45.

If the system has no paths or cuts of size 1, then the function $\Psi_0(p)$ is S-shaped; that is, there exists a p-value p_0 between 0 and 1, such that $\Psi_0(p) \leq p$ for $p \in [0, p_0]$ and $\Psi(p) \geq p$ for $p \in [p_0, 1]$. Consider, for example, the reliability function for a series connection of two parallel subsystems, each having two components. For this system, $\Psi_0(p) = [2p - p^2]^2$. It is easy to verify that it has an S-shaped form. An S-shaped function has always an inflexion point, and its second derivative *changes its sign*.

Now let us use the following equality:

$$\frac{d\Psi_0(p)}{dp} = \sum_{i=1}^{n} \text{Imp}_i(p). \tag{4.6.2}$$

In words: if all components have equal reliability then the derivative of system reliability function $R = \Psi_0(p)$ with respect to p is equal to the sum of all component importance functions. (The proof follows from the chain rule of differentiation after setting $p_i \equiv p$.) The solution follows if you differentiate both sides of the last equality with respect to p.

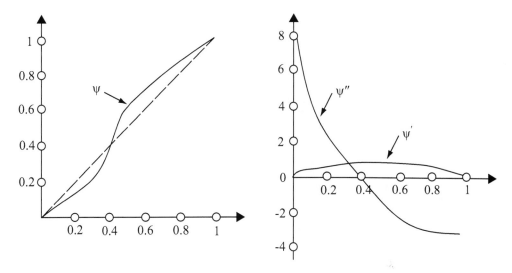

Figure 4.5: S-shaped function $\psi(p)$ and its first and the second derivative

10. The system is a series connection of n components with reliability $p_1 < p_2 < ... < p_n$. Prove that the largest BIM has the less reliable component. What is the most important component in parallel system?

11. System has five components. Components 2 and 3 are in series. Components 4 and 5 are in series too. Subsystems $(2,3)$ and $(4,5)$ are in parallel. The system of these four components is in a series connection with component 1. All components are independent. Component i has two states, *up* and *down*, denoted as $x_i = 1$ or $x_i = 0$, respectively, $i = 1, ..., 5$.

a). Find system structure function $\phi(\mathbf{x}) = \phi(x_1, x_2, ..., x_5)$.

b). Find system reliability assuming that component i is *up* with probability p_i.

c). Assuming that all $p_i = p$, find system reliability function and component importance.

12. For the system described in Problem 11, assume that component i has lifetime $\tau_i \sim Exp(\lambda_i)$. Find the expression for system reliability $R_s(t)$.

13. A radar system consists of three identical independently operating stations. The system is *UP* if at least two stations are *up*. Assume that station lifetime $\tau \sim Exp(\lambda)$. After the system enters the *DOWN* state, it is repaired and brought to its initial state. The repair lasts time $1/\lambda$. Find the mean duration of the *UP* period and system availability.

14. A system consists of three units A, B, and C. If only one unit is *up*, the system is *UP*. The same is true for two units. But if all three units are *up*, the system is *DOWN* because the power supply for all units fails: there is not enough power to support all three units. Is this system monotone?

15. Consider a small channel-type network. Source node s is connected to "first stage" nodes a and b. Each of these two nodes is connected to the "second stage" nodes c and d. Each of these nodes is connected to the terminal node t. Edges (s, a) and (s, b) have numbers 1 and 2. Edges (a, c) and (a, d) have numbers 3 and 4. Edges (b, c) and (b, d) have numbers 5 and 6, respectively. Edges (c, t) and (d, t) have numbers 7 and 8. Edge i is up with probability p_i, all edges are independent. The network is *UP* if s and t are connected.

Design an algorithm for simulating the network reliability.

16. Solve Examples 4.2.1 and 4.2.2.

17. Suppose, our system is not monotone. Is it true that $P(\tau_s > t_0) = R_s(t_0)$, where R_s is given by (4.2.2)?

18. Check (4.5.4) and (4.5.5).

Chapter 5

Border States and Reliability Gradient

> Crossing the border may be a one-way street.
>
> Chinese saying.

5.1 Definition of Border States

In this chapter we introduce the so-called system *border states* and demonstrate that their probability is intimately related to the system reliability gradient function. This paves the way for calculating the reliability gradient via the probabilities of the border states. The corresponding MC algorithm will be described later, in Chapter 9.

Let us consider a monotone system of n binary components. All 2^n binary states are divided into two classes: UP and $DOWN$.

Definition 5.1.1

System state $\mathbf{w} = (w_1, ..., w_n) \in DOWN$ is called *direct neighbor* or simply *neighbor* of state $\mathbf{e} \in UP$ if $\mathbf{w}$ differs from $\mathbf{e}$ in exactly one position.#

For example, consider so-called two-out-of-three system which has three components and is UP, by definition, if at least two of them are UP. So,

$$DN = \{(0,0,0), (0,0,1), (0,1,0), (1,0,0)\}$$
$$UP = \{(1,1,0), (1,0,1), (0,1,1), (1,1,1)\}.$$

Thus the state $(0,0,1)$ is the border state of $(1,0,1) \in UP$ and also the border state of $(0,1,1) \in UP$. Similarly, $(0,1,0)$ and $(1,0,0)$ are border states of (1,1,0).

The set of all neighbor states is called *border set* and denoted as DN^*. Obviously, $DN^* \subseteq DOWN$.

It turns out that the reliability gradient vector introduced in Chapter 4 is closely related to the border states. To reveal this connection, we introduce an *evolution process* on system components.

Assume that at $t = 0$ each component is *down*. Component i is born after random time $\tau_i \sim Exp(\lambda_i)$. After the "birth", component remains *up* "forever". Note that for fixed time t_0,

$$P(\tau_i \leq t_0) = P(\text{ component } i \text{ is } up \text{ at } t_0) = p_i = 1 - e^{-\lambda_i t_0}. \quad (5.1.1)$$

Example 5.1.1. Evolution of a three component system.

Fig. 5.1 below illustrates the evolution of three component two-out-of-three system.

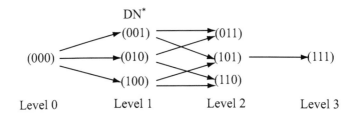

Figure 5.1: All states on Level 1 are border states

After the birth of the first component, the system moves from state (0,0,0) into one of the states on level 1, afterwards - jumps to a state on level 2 and later - to level 3. We consider the states on levels 2 and 3 as one "large" absorbing state. All these states constitute the system's UP state. All other states are defined as the $DOWN$ states of the system. The directions of possible transitions are shown by arrows. Note that the states directly before the UP states, on level 1, are the *border* states. They reach UP in one jump, in a single transition.#

Consider two system states $\mathbf{v} = (v_1, v_2, ..., v_{i-1}, 0, v_{i+1}, ..., v_n)$ and $\mathbf{w} = (v_1, v_2, ..., v_{i-1}, 1, v_{i+1}, ..., v_n)$. Suppose that at time t the system is in state $\mathbf{v}$. What is the probability that during a small time interval δt the system will move from $\mathbf{v}$ to $\mathbf{w}$? Obviously, it will happen if and only if the component i is born during this interval, and all other components which are in state 0 will not become alive during the same interval. The first event has probability $\lambda_i \cdot \delta t + o(\delta t)$ as $\delta t \to 0$ (because of the exponentially distributed lifetime, see Chapter 3). The second event has probability

$$\prod_{j \neq i}(1 - O(\lambda_j \cdot \delta t)) = 1 - O(\delta t).$$

Then the probability that during $[t, t + \delta t]$ there will be the transition $\mathbf{v} \to \mathbf{w}$ is $\lambda_i \delta t + o(\delta t)$.

Let $\mathbf{v}$ be a border state of the system, i.e. $\mathbf{v} \in DN^*$. Denote by $\Gamma(\mathbf{v})$ the sum of λ_i over all set of indices i such that $\mathbf{v} + (0, ..., 1_i, 0, ..., 0) \in UP$. Call $\Gamma(\mathbf{v})$ the *flow* from $\mathbf{v}$ into UP. Formally

$$\Gamma(\mathbf{v}) = \sum_{[\mathbf{v} \in DN^*, \mathbf{v}+(0,...,0,1_i,0,...,0) \in UP]} \lambda_i. \tag{5.1.2}$$

Remark. Let $\mathbf{v} \in DN^*$. *Not all* down elements in $\mathbf{v}$ have the property that being turned into *up* they cause the system to become UP. Consider, for example a bridge structure with edges $e_1 = (s, 1), e_2 = (s, 2), e_3 = (1, 2), e_4 = (1, t), e_5 = (2, t)$ and $\mathbf{v} = (1, 1, 0, 0, 0)$. When edge e_3 becomes *up*, the system remains in $DOWN$. In this example $\Gamma(\mathbf{v}) = (\lambda_4 + \lambda_5).\#$

5.2 Gradient and Border States

We need two other notations. Let $R(p_1(t), ..., p_n(t))$ be the probability that the system is UP at the instant t. Let $P(\mathbf{v}; t)$ be the probability that the system is in state $\mathbf{v}$ at time t. Now let us consider the event "the system is in UP at time $t + \delta t$". This event takes place if at time t the system was already in the UP set or at time t it was in one of the border states and went during this interval from a border state to UP. All other possibilities which involve more than one transition during $[t, t + \delta t]$ have probability $o(\delta t)$.

Formally,

$$R(p_1(t + \delta t), ..., p_n(t + \delta t)) = R(p_1(t), ..., p_n(t)) + \tag{5.2.1}$$
$$\sum_{\mathbf{v} \in DN^*} P(\mathbf{v}; t)\Gamma(\mathbf{v}) \cdot \delta t + o(\delta t).$$

Transfer $R(p_1(t), ..., p_n(t))$ to the left-hand side, divide both sides of (5.2.1) by δt and let $\delta t \to 0$. We arrive at the following relationship:

$$\frac{dR(p_1(t), ..., p_n(t))}{dt} = \sum_{\mathbf{v} \in DN^*} P(\mathbf{v}; t)\Gamma(\mathbf{v}). \tag{5.2.2}$$

Now, represent the left-hand side of (5.2.2) in an alternative form using the chain rule of differentiation:

$$\frac{dR(p_1(t), ..., p_n(t))}{dt} = \sum_{j=1}^{n} \frac{\partial R}{\partial p_j} \cdot \frac{dp_j(t)}{dt} = \qquad (5.2.3)$$

$$|p_j(t) = 1 - e^{-\lambda_j t}, q_j = e^{-\lambda_j t}| = \sum_{j=1}^{n} \frac{\partial R}{\partial p_j} q_j \cdot \lambda_j$$

$$= \nabla R \bullet \{q_1 \lambda_1, ..., q_n \lambda_n\}.$$

We use in (5.2.3) the shorthand notation for the vector scalar product: if $\mathbf{x} = (x_1, x_2, ..., x_n)$ and $\mathbf{y} = (y_1, y_2, ..., y_n)$,
    $\mathbf{x} \bullet \mathbf{y} = \sum_{i=1}^{n} x_i y_i$. We denote $\nabla R = (\partial R / \partial p_1, ..., \partial R / \partial p_n)$.

Comparing (5.2.2) and (5.2.3) we arrive at the desired relationship between the gradient vector and the border state probabilities:

$$\nabla R \bullet \{q_1 \lambda_1, ..., q_n \lambda_n\} = \sum_{\mathbf{v} \in DN^*} P(\mathbf{v}; t) \Gamma(\mathbf{v}). \qquad (5.2.4)$$

We see therefore that components of the reliability gradient vector can be expressed via the probabilities of system border states. In the situation when the system reliability function $R(p_1(t), ..., p_n(t))$ is not available in explicit analytic form, (5.2.4) opens an alternative way of calculating the components of ∇R. This becomes possible in a form of a Monte Carlo procedure by using properties of so-called Lomonosov's algorithm which will be introduced in Chapter 9.

Example 5.1.1 - continued.
Let us check the last formula for the three component system considered in Example 5.1.1. We will omit the argument t and denote by p_i the probability that the component i is *up*.

Summing all probabilities of the system UP state (four states) we obtain the following expression:

$$R_s = (1 - p_1)p_2 p_3 + (1 - p_2)p_1 p_3 + (1 - p_3)p_1 p_2 + p_1 p_2 p_3. \qquad (5.2.5)$$

Now it is easy to obtain the partial derivatives:
$\frac{\partial R}{\partial p_1} = (1 - p_2)p_3 + p_2(1 - p_3).$

$\frac{\partial R}{\partial p_2} = (1 - p_1)p_3 + p_1(1 - p_3).$

$\frac{\partial R}{\partial p_3} = (1 - p_1)p_2 + p_1(1 - p_2).$

Let us write the expression for the right-hand side of (5.2.4). It is easy to verify (see Fig. 5.1) that in our example

$$\sum_{\mathbf{v} \in DN^*} P(v; t) \Gamma(\mathbf{v}) = (1 - p_1)(1 - p_2)p_3 \cdot (\lambda_1 + \lambda_2) + \quad (5.2.6)$$
$$(1 - p_1)p_2(1 - p_3) \cdot (\lambda_1 + \lambda_3) + p_1(1 - p_2)(1 - p_3) \cdot (\lambda_2 + \lambda_3).$$

Now let us check whether the relationship (5.2.4) holds true. For this purpose, let us compare the coefficients at λ_1. In the expression (5.2.6) this coefficient is

$$(1 - p_1)[(1 - p_2)p_3 + p_2(1 - p_3)].$$

But it is exactly the same as the first coordinate of the gradient vector times q_1, i.e. $\partial R / \partial p_1$ multiplied by $q_1 = 1 - p_1$. Similarly, we can check the coefficients at λ_2 and λ_3. So, the formula (5.2.4) works!#

Remark. We have considered in this and in the previous chapter various events related to transitions from state to state. To be more specific, let us consider a system of two independent components with exponentially distributed lifetimes, with parameters λ_1 and λ_2. Let us look more closely at three system states (0,0) - both components are *down*, (1,0) and (0,1) (the first component is *up*, the second is *down*, and vice versa). We proved in Chapter 3 that the transition $(0, 0) \rightarrow (1, 0)$ takes place with probability $\lambda_1/(\lambda_1 + \lambda_2)$. On the other hand we considered the probability of the transition from state (0,0) into (1,0) during a small time interval $[t, t + \delta t]$. The result was $\lambda_1 \delta t + o(\delta t)$. So, it seems that we have two different results for the same event. But in fact, these are quite *different* events.

The first one is the probability that the system eventually moves from state (0,0) into state (1,0), without restricting the time during which this transition will take place. If we consider all histories describing the transition from (0,0) into (1,0) or (0,1), then the probability $\lambda_1/(\lambda_1 + \lambda_2)$ is the relative weight of all trajectories of type $(0, 0) \rightarrow (1, 0)$ among all trajectories leading from $(0, 0)$ to the UP set.

The second situation dealt with system evolution during small time interval $[t, t + \delta t]$. If the system is in state (0,0) at time t, then at time $t + \delta t$ it will be in (1,0) with probability $\lambda_1 \delta t + o(\delta t)$, or in state (0,1) with probability $\lambda_2 \delta t + o(\delta t)$, or remain in state (0,0) with complementary probability $1 - (\lambda_1 + \lambda_2)\delta t + o(\delta t)$.#

5.3   Problems and Exercises

1. Suppose that the right-hand side of (5.2.4) has the following form:

$$\psi_1 \cdot (\lambda_1 + \lambda_2) + \psi_2 \cdot (\lambda_2 + \lambda_3) + \psi_3 \cdot (\lambda_1 + \lambda_3),$$

where ψ_i are known functions of p_1, p_2, p_3. Find the expression for the gradient vector.

2. A network has nodes a, b, c and edges $(a, b), (b, c), (a, c)$. Edge reliability is p_1, p_2, p_3, respectively. Nodes are not subject to failures. The network is *UP* if it is connected. Find the expression for edge importance. Suppose $p_1 < p_2 < p_3$, all $p_i > 1/2$. Which edge is the most and the less important?

3. A system consists of a series connection of one component and a block of three components in parallel, se Fig. 5.2. Find all border states.

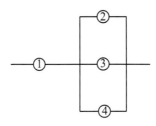

Figure 5.2: Component i has reliability p_i, all components are independent

4. Assume in Problem 3 that all $p_i = p$. Find the most important component.

5. Consider a four-node network with nodes s, a, t, b. It has edges $(s, a), (a, t), (t, b), (s, b)$ and (s, t). The network is *UP* if s and t are connected. Find out all border states for this network with exactly *two up* edges.

6. Check (5.2.4) for the system shown on Fig. 5.2.

7. Suppose that in the network of Problem 5 each edge has small failure probability α. Use the definition of component importance given by formula (4.5.7). Compute network reliability using the B-P approximation. Prove that (s, t) is the most important component.

8. Consider a series connection of two parallel systems, one with three components, another with two. Assume that all components have equal failure probabilities, all components fail independently. Compute the BIM for system components. Show that for all values of reliability p, the components of the first system have *smaller* BIM's than the BIM's of the second.

Chapter 6

Order Statistics and Destruction Spectrum

6.1 Reminder of Basics in Order Statistics

The best source on order statistics is Herbert A. David's book [9].

Let $X_1, X_2, ..., X_n$ be i.i.d. random variables, $X_i \sim F(t)$. We call the collection $\{X_1, X_2, ..., X_n\}$ a random sample. Let $X_{(1)}$ denote the smallest value in the random sample, let $X_{(2)}$ denote the next smallest value in the sample, and so on. In this way, $X_{(n)}$ denotes the largest value in the sample. In other words,

$$X_{(1)} \leq X_{(2)} \leq \ldots \leq X_{(n)}. \tag{6.1.1}$$

Definition 6.1.1. $X_{(r)}$ is called the r-th *order statistic*, $r = 1, 2, ..., n.\#$

Often the notation $X_{(r)}^{(n)}$ is used to denote the r-th order statistic in a sample of size n.

A good way to get used to the notion of the r-th order statistic is to imagine the following generator of sample values x_r of X_r. Generate a sample of size n from the population with CDF $F(t)$, $X \sim F(t)$. Denote these sample values by $x_1, x_2, ..., x_n$. Now *order* these values in increasing order. Pick up the r-th smallest value. Repeat this procedure k times. In this way we obtain k sample values of $X_{(r)}$.

We assume further that all X_i are nonnegative and continuous r.v.'s, and that X_i represents the lifetime of component i, $i = 1, 2, ..., n$.

Claim 6.1.1

$$F_{(r)}(t) = P(X_{(r)} \leq t) = P(\text{at least } r \text{ out of } n \; X_i\text{-s are } \leq t) = \quad (6.1.2)$$

$$\sum_{j=r}^{n} C_n^j [F(t)]^j [1 - F(t)]^{n-j}. \#$$

Here C_n^j is the number of combinations of j out of n:

$$C_n^j = \frac{n!}{j!(n-j)!}$$

As an exercise, derive (6.1.2) by probabilistic arguments.

Order statistics $X_{(1)} = X_{min}$ and $X_{(n)} = X_{max}$ are of particular interest to reliability theory. A system of n independent components which fails if one of the components has failed is called *series* system. Obviously, if all component lifetimes have the same CDF $F(t)$, series system lifetime is determined by the expression

$$X_{min} \sim F_{min}(t) = 1 - [1 - F(t)]^n. \qquad (6.1.3)$$

As an exercise, obtain this result from (6.1.2) by setting $r = 1$.

For the case of non identically distributed lifetimes, the last formula is slightly more complex:

$$X_{min} \sim F_{min}(t) = 1 - \prod_{i=1}^{n} [1 - F_i(t)]. \qquad (6.1.4)$$

As an exercise, verify formulas (6.1.3) and (6.1.4) by using probabilistic reasoning.

A system of n independent components which fails if **all** its components have failed is called *parallel* system. Obviously, in parallel system CDF for identically distributed lifetimes is determined by the expression

$$X_{max} \sim F_{max}(t) = [F(t)]^n. \qquad (6.1.5)$$

Verify that this formula follows from (6.1.2) by setting $r = n$.

For the case of non identically distributed lifetimes, the last formula is slightly more complex:

$$X_{max} \sim F_{max}(t) = \prod_{i=1}^{n} F_i(t). \qquad (6.1.6)$$

As an exercise, verify formulas (6.1.5) and (6.1.6) by using probabilistic reasoning.

Remark. Consider a system of 5 independent components with identically distributed lifetimes. The system is called 3-out-of-5, if it is *UP* if at least 3 out of 5 components are *up*. System lifetime is, therefore, described by the third order statistic in a sample of 5. There are many technical systems which consist of n similar components and are operational (*UP*) as long as at least k of them are operational. Such systems are called *k-out-of-n* systems, see e.g. [19].#

6.2 Min-Max Calculus for Series-Parallel Systems

Systems studied in reliability theory can be sometimes represented as a series connection of several subsystems, each of which is a parallel system, or as a parallel connection of subsystems each one being a series system. Fig. 6.1 below illustrates these types of systems.

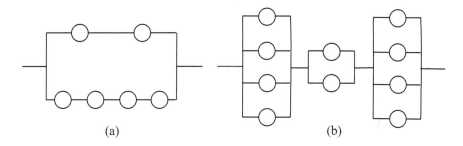

(a)   (b)

Figure 6.1: (a) parallel-series system; (b) series-parallel system

There is an extremely simple method of finding the analytic form of system lifetime in case of series-parallel or parallel-series system. (All components are assumed to be independent.)

Let us demonstrate it by an example.

Let us consider a system of seven components shown on Fig. 6.2. The system is *UP* if there is connection between s and t.

The subsystem of components 1,2 has lifetime $\tau_{12} = \min(\tau_1, \tau_2)$. Similarly, $\tau_{34} = \min(\tau_3, \tau_4)$. Therefore, the block of first four components has lifetime $\tau_{1234} = \max(\tau_{12}, \tau_{3,4})$. Similarly, $\tau_{567} = \max(\tau_5, \tau_6, \tau_7)$. Finally, $\tau_{system} = \min(\tau_{1234}, \tau_{567})$.

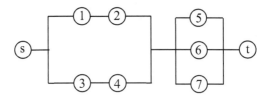

Figure 6.2: Series-parallel system of 7 components

6.3   Destruction Spectrum (D-spectrum)

In this section we will consider networks with reliable nodes and unreliable edges. All edges have identically distributed lifetimes.

Consider the system pictured on Fig. 6.3.

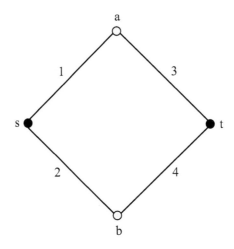

Figure 6.3: A system of four components

It consists of four identical and independent components. Edge lifetime $\tau \sim F(t)$. At $t = 0$ all components are up. The system fails at the instant of the loss of $s - t$ connection. For example, if component 1 fails at $t = 5$, component 3 at time $t = 7.5$, and component 2 at $t = 9$, then the disconnection appears at the instant $t = 9$, i.e. at the instant of the *third* failure.

Denote by f_r the probability of the event A_r="system fails at the instant of the r-th failure". In other words, A_r takes place if and only if the system survives the first $r - 1$ component failures and fails at the instant of r-th failure.

Definition 6.3.1. The set of probabilities

$$Sp = [f_1, f_2, ..., f_n] \tag{6.3.1}$$

is called the *destruction spectrum*, or simply D-spectrum.#

The spectrum depends on system structure and failure definition. In the above case the system can fail on the second or on the third edge failure. If the system failure is defined as a loss of connectivity, the failure will always appear at the second edge failure, no matter in which order the edges fail. For example, if edge 1 and 3 fail, the node a becomes disconnected. If first edges 1 and 4 fail, nodes s, b become disconnected from a, t.

Note also that spectrum is a *distribution*, i.e. $\sum_{i=1}^{n} f_r = 1$.

Note also (and this is very important!) that the spectrum **does not depend on edge lifetime distribution**.

Let us find the destruction spectrum for the system on Fig. 6.3. The system never fails at the first failure and always fails after three edges are down. So, we must find only f_2 and f_3. There are 4!=24 different and equally probable orderings of edge failures. It is easy to find out by direct enumeration that $f_3 = 8/24 = 1/3$. Therefore $Sp = [0, 2/3, 1/3, 0]$. (Check it!)

Now comes an important turn: we connect the system lifetime with the spectrum and order statistics.

Suppose edge lifetime is $\tau \sim F(t)$.

If the system fails with the second failure (the event A_2), system lifetime coincides with $X_{(2)}$ and is distributed as the second order statistic from sample of 4, i.e. has CDF $F_{(2)}(t)$ for sample size $n = 4$. Similarly, if the event A_3 takes place, system lifetime is distributed as $F_{(3)}(t)$. Therefore, by the total probability formula,

$$P(\tau_s \le t) = F_s(t) = f_2 F_{(2)}(t) + f_3 F_{(3)}(t). \tag{6.3.2}$$

As an exercise, consider the system on Fig. 6.3. Let t be fixed. Denote $q = F(t), p = 1 - F(t)$ and find the explicit form of $F_s(t)$.

Hint. Note that the system on Fig. 6.3 is a parallel connection of two series systems. An alternative approach is to use system spectrum which is $Sp = [0, 2/3, 1/3, 0]$ and (6.3.2). Of course, both answers must coincide.

It is easy to generalize (6.3.2) to the case of n components.

Claim 6.3.1

If (i) edge lifetimes are i.i.d. r.v.'s $X_i \sim F(t)$ and (ii) the destruction spectrum is $Sp = [f_1, f_2, ..., f_n]$, then the network lifetime τ_s has the following CDF:

$$P(\tau_s \leq t) = F_s(t) = \sum_{r=1}^{n} f_r F_{(r)}(t). \# \qquad (6.3.3)$$

As an exercise prove this claim.

Corollary 6.3.1

Denote $P(\tau_s > t) = R_s = 1 - F_s(t)$, and replace in the expressions for the order statistics $F_{(r)}(t)$ and $1 - F_{(r)}(t)$ by $1 - p$ and p, respectively, for all $r = 1, ., n$.

Then the expression for R_s in this new notation equals the system static reliability, in which p is the static probability that the component is *up*.

See also Sections 4.1, 4.2.#

Remark 1. Suppose that the unreliable elements in the network are the nodes and not the edges. In this situation the D-spectrum may be defined for network nodes. For example, consider the network on Fig. 6.3 and suppose that all edges and nodes s, t (the terminals) are not subject to failure and the two remaining nodes are. Obviously, the network fails if and only if both these nodes fail. Thus the *node D-spectrum* has the form $[f_1 = 0, f_2 = 1]$.#

Remark 2. Suppose that at $t = 0$ all network edges are *down*. Edge e *is born* at random time instant $X_e \sim F(t)$. Let us check the system state after each birth. There is such number k having the property that after the first $k - 1$ births the system was *DOWN* but at the instant of the k-th birth it becomes *UP* (and remains *UP* "forever"). Denote by b_k the probability of this event. Obviously, the collection $S_c = [b_1, b_2, ..., b_n]$ is a distribution and it will be called the *construction spectrum* or simply C-spectrum. D-spectrum and C-spectrum are closely connected, see Problem 9 in this chapter.#

6.4   Formula for the Number of Minimal size Min-Cuts

A cut is a set of components whose failure causes the failure of the whole system. Minimal (or min) cut is a cut which does not contain proper subsets which are also cuts. For example, components (1,2,3) constitute a cut for the system shown on Fig. 6.3. This cut, however, is not minimal: it contains smaller subsets $(1,2)$ and $(2,3)$ which are also cuts. $(1,2)$ is a minimal cut because it cannot be reduced further and remain a cut. We will be interested in estimating the number of min-cuts of *minimal* size.

Suppose that the *first nonzero* element in the D - spectrum is f_r. Then

$$P(\tau_s \le t) = F_s(t) = f_r F_{(r)}(t) + f_{r+1} F_{(r+1)}(t) + \dots \qquad (6.4.1)$$

Suppose that $F(t) = 1 - e^{-\lambda t}$, t is fixed, and $\lambda \to 0$. Then, $[F(t)]^r = \lambda^r t^r + o(\lambda^r)$. Then, according to (6.1.2), the CDF of the order statistics can be written for $\lambda \to 0$ as

$$F_{(r)}(t) = C_n^r (\lambda t)^r + o(\lambda^r)$$

because all $F_{(i)}(t)$ for $i > r$ are of order $o(\lambda^r)$. After little algebra, we obtain that

$$F_s(t) = f_r C_n^r \lambda^r t^r + o(\lambda^r). \qquad (6.4.2)$$

On the other hand, the main term of the system failure probability in the Burtin-Pittel approximation (see Chapter 4) equals

$$W \cdot \lambda^r t^r$$

where W is the number of minimal cut-sets of minimal size r. Comparing this expression with the right-hand side of (6.4.2), we arrive at the following simple formula:

$$W = f_r \frac{n!}{r!(n-r)!}. \qquad (6.4.3)$$

Thus, knowing the first nonzero term of the spectrum, we can easily calculate the number of min size min-cuts.

For example, for the network shown on Fig. 6.3, we have $f_2 = 2/3$, $n = 4, r = 2$. (6.4.3) gives the result $W = 4$. Indeed, there are four cuts of size two: $(1, 2), (3, 4), (1, 4)$, and $(3, 2)$. More examples will be presented in Chapter 8.

In conclusion, let us note that we found an important *topological* parameter of a network with given operational criterion: the min-cut size and the number of such cuts. If a reliable network fails, most probably the failure will be in one of its "soft" spots, the minimal size min-cuts. The analytic form of the edge lifetime CDF is not essential. Important is only the assumption that all edge sequences of failure appearance are equally probable.

6.5   Problems and Exercises

1. Find the spectrum of the bridge-type system ($n = 5$ edges). Failure is defined as loss of the $s - t$ connectivity.

2. Suppose $X_1, ..., X_{10}$ are i.i.d., $X_i \sim Exp(1)$. Design an algorithm for generating a sample of size $m = 15$ from the population of $X_{(7)}$, i.e. from the population of the seventh order statistics.

3. Derive (6.1.2), (6.1.3), (6.1.5), and (6.1.6) by using probabilistic arguments.

4. Solve exercises in Section 6.3 using spectrum.

5. Prove Claim 6.3.1.

6. Suppose that in your network all edges are *down* at $t = 0$. Edges have "birth" after random time $\tau \sim F(t)$. After the birth, each edge becomes permanently *up*. Edges are born independently, and their birth times are i.i.d. r.v.'s.

Similarly to the destruction spectrum (call it D-spectrum), develop an analogous theory for the "construction" spectrum (call it **C-spectrum**).

Define the network birth-time as the moment at which the network becomes UP. Formulate a claim similar to Claim 6.3.1 for the network birth-time distribution.

7.   3-out-of-5 system. Five independent experts start simultaneously an evaluation process of a bridge project. The evaluation time for expert i is X_i. $X_i, i = 1, 2, 3, 4, 5$ are assumed to be i.i.d. random variables. At the moment when *three* experts have finished their reports, the whole group terminates its work. Suppose that $X_i \sim F(x)$. Find out the distribution function of the time interval needed for the evaluation.

8. A pentagon-type network has nodes a, b, c, d, e. Nodes (a, d) are the terminals, other nodes are subject to failure. The network has edges $1 = (a, b)$, $2 = (b, c)$, $3 = (c, d)$, $4 = (d, e)$, $5 = (e, a)$. Edges do not fail. Nodes b, c, e are subject to failures. Find the network D-spectrum for the terminal connectivity criterion.

9. Find the C-spectrum for the $s - t$ network on Fig. 6.3. Try to establish the connection between the D-spectrum and the C-spectrum for the same network.

10. System consists of seven identical and independent components. The first three are organized into a 2-out-of-3 block. The remaining four - into a 2-out-of-4 block. These two blocks are connected in series. Component lifetime $\tau_i \sim F(t)$. Derive a formula for system lifetime distribution $F_s(t)$.

Chapter 7

Monte Carlo Estimation of Convolutions

This chapter can be omitted during the first reading of the book. Convolutions appear in our exposition only in the connection with so-called "Turnip algorithm" for network reliability estimation, in Chapter 9. This algorithm asks for calculating a CDF of a sum of r independent exponentially distributed random variables, see (9.2.7). If $r < 30 - 40$, we can rely on analytic expression presented in Appendix B, see formula (5) there. For larger number of r (this will be the case of relatively large networks) the analytic calculations become unstable and may produce large computational error. This happens because the above formula is constructed as a sum of very large quantities, close to each other by their absolute values, with alternating signs. It turns out that it is much safer to use Monte Carlo simulation to estimate the convolutions.

From the point of view of simulation technique, our algorithm for estimating convolutions is an interesting example of using importance sampling to increase considerably an algorithm efficiency in comparison to a crude Monte Carlo.

7.1 Model Formulation. Using Crude Monte Carlo for Calculating Convolutions

Let $\tau_i \sim F_i(t), f_i(t), i = 1, 2, ..., m$, be positive and independent random variables. Our purpose is to find the CDF of their sum $T_m = \tau_1 + ... + \tau_m$.

There are two cases when the CDF of T_m can be easily found in a closed form: $F_i(t)$ - normal (with negligible negative tails) and $F_i(t) \sim Gamma(k_i, \lambda)$. Otherwise, we must use an analytic approach which is quite inconvenient for numerical calculations. An alternative is to use *simulation*.

The CDF of the sum of independent random variables is called *convolution*. Our goal is to estimate $P(T_m \le T) = F^{(m)}(T)$.

What first comes to mind is to calculate convolutions using the Crude Monte Carlo (CMC) method, which works as follows:

Algorithm 7.1 - CMC for Calculating Convolutions.
1. **Set NUM=0.**
2. **Simulate** $\tau_i = t_i$, for $i = 1, 2., , , .m$.
3. **If** $\sum_{i=1}^{m} t_i \le T$, **set** $I = 1$, **else set** $I = 0$.
4. **Put** NUM:=NUM +I.
5. **Repeat** 2-4 N times, **calculate**

$$\hat{F}^{(m)}(T) = \frac{NUM}{N}. \tag{7.1.1}$$

Obviously,

$$E[\hat{F}^{(m)}(T)] = F^{(m)}(T), \tag{7.1.2}$$

and thus the CMC estimator is unbiased. It is easy to verify (do it!) that

$$Var[\hat{F}^{(m)}(T)] = F^{(m)}(T)(1 - F^{(m)}(T))/N, \tag{7.1.3}$$

and therefore the relative error of the CMC is unbounded as $F^{(m)}(T) \to 0$.

7.2   Analytic Approach

By definition, $F^{(1)}(T) \equiv F_1(T)$,

$$F^{(2)}(T) = \int_0^T F^{(1)}(T - y)f_2(y)dy = P(\tau_1 + \tau_2 \le T), ..., \tag{7.2.1}$$

$$F^{(m)}(T) = \int_0^T F^{(m-1)}(T - y)f_m(y)dy = P(\tau_1 + ... + \tau_m \le T).$$

These formulas have a very transparent probabilistic interpretation. Let us consider, for example, $F^{(2)}(T) = P(\tau_1 + \tau_2 \le T)$. How can sum of τ_1 and τ_2 be less or equal to T? One random variable, e.g. τ_2 equals y, and another is less or equal $T - y$. The corresponding probabilities are multiplied since both r.v.'s are independent, and the sum of all such possibilities is the corresponding integral over $[0, T]$.

Let us look at the convolution formulas (7.2.1) from another angle, namely how are they related to multiple integrals. By the definition, and using the representation of multiple integral via repeated integral, we can write that

$$F^{(2)}(T) = \int_0^T F^{(1)}(T - y) f_2(y) dy = \tag{7.2.2}$$

$$\int_{y_2:0}^T f_2(y_2) \left(\int_{y_1:0}^{T-y_2} f_1(y_1) dy_1 \right) dy_2 =$$

$$\int \int_{\{0 \leq y_1 + y_2 \leq T\}} f_1(y_1) f_2(y_2) dy_1 dy_2.$$

Thus, $F^{(2)}(t)$ is represented as a double integral over a triangular area $0 \leq y_1 + y_2 \leq T, y_1 \geq 0, y_2 \geq 0$, see Fig 7.1.

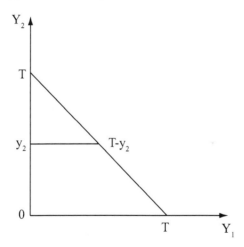

Figure 7.1: Integration area for double integral

In a similar way, one can represent $F^{(3)}(T)$ as a triple integral of $f_1(y_1) \cdot f_2(y_2) \cdot f_3(y_3)$ over triangular area (also called simplex) $\{0 \leq y_1 + y_2 + y_3 \leq T, y_i \geq 0\}$. It can be shown that

$$F^{(m)}(T) = \int (m) \int_{\sum_{i=1}^m y_i \leq T, y_i \geq 0} \prod_{i=1}^m f_i(y_i) dy_1 \dots dy_m. \tag{7.2.3}$$

We are interested in finding an unbiased estimator of $F^{(m)}(T)$. Since in most applications this quantity is small, the CMC algorithm described above is non efficient and has a very large relative error. To overcome this obstacle we propose a new algorithm described below, see [21].

7.3   Conditional Densities and Modified Algorithm

Let $\tau_i, i = 1, ..., m$ be nonnegative i.r.v.'s with CDF $F_i(t)$ and density functions $f_i(t)$. Consider a random vector

$$\mathbf{X} = (X_1, X_2, ..., X_{m-1}) \tag{7.3.1}$$

with the joint density $f_\mathbf{X}(v_1, ..., v_{m-1})$, defined as follows:

$$f_\mathbf{X}(v_1, ..., v_{m-1}) = f_{X_1}(v_1) f_{X_2|X_1}(v_2|v_1) \cdots \tag{7.3.2}$$
$$f_{X_{m-1}|X_1,...,X_{m-2}}(v_{m-1}|v_1, ..., v_{m-2}),$$

where

$$f_{X_1}(v_1) = \frac{f_1(v_1)}{F_1(T)}, \tag{7.3.3}$$

and

$$f_{X_j|X_1,...,X_{j-1}}(v_j|v_1, ..., v_{j-1}) = \frac{f_j(v_j)}{F_j(T - v_1 - ... - v_{j-1})}, \tag{7.3.4}$$

$v_j \in [0, T - v_1 - ... - v_{j-1}]$, for $j = 2, ..., m - 1$. Define the following r.v.:

$$B_m(T) = F_1(T) \cdot F_2(T - X_1) \cdot F_3(T - X_1 - X_2) \cdot ... \tag{7.3.5}$$
$$\cdot F_m(T - X_1 - ... - X_{m-1}).$$

Now we are ready to formulate our main result.

Claim 7.3.1

$$E[B_m(T)] = \int (m) \int_{\{v_i \geq 0, v_1 + ... v_m \leq T\}} \prod_{i=1}^{m} f_i(v_i) dv_1 ... dv_m. \tag{7.3.6}$$

Proof.

The proof is not difficult. We shall demonstrate it for $m = 3$.

$$B_3(T) = F_1(T) F_2(T - X_1) F_3(T - X_1 - X_2). \tag{7.3.7}$$

Then

$$E[B_3(T)] = \qquad (7.3.8)$$

$$\int\int_{\{v_1\geq 0, v_2\geq 0, v_1+v_2\leq T\}} F_1(T)F_2(T-v_1)F_3(T-v_1-v_2)\cdot$$

$$\cdot\frac{f_1(v_1)f_2(v_2)}{F_1(T)F_2(T-v_1)}dv_1dv_2.$$

After simple algebra we obtain that

$$E[B_3(T)] = \quad (7.3.9)$$

$$\int\int_{\{v_1\geq 0, v_2\geq 0, v_1+v_2\leq T\}} f_1(v_1)f_2(v_2)\left(\int_0^{T-v_1-v_2} f_3(x)dx\right)dv_1dv_2.$$

And the last expression is the integral of $f_1(v_1)f_2(v_2)f_3(v_3)$ over the three dimensional simplex $\{v_i \geq 0, v_1 + v_2 + v_3 \leq T\}.\#$

7.4 Generating $B_m(T)$

1. Generate $\mathbf{X} = (X_1, ..., X_{m-1})$ recursively: Generate $X_1 = x_1$ from the population with d.f. $f_{X_1}(\cdot)$. Use x_1 to generate the value x_2 of r.v. X_2 from the conditional density $f_{X_2|X_1}(\cdot)$, etc.
2. Substitute $\mathbf{X}=\mathbf{x} = (x_1, x_2, ..., x_{m-1})$ into (7.3.5).

For the particular case of $\tau_i \sim Exp(\lambda_i)$ the generation procedure is as follows. Let $\xi \sim U(0, 1)$.

Generate $\xi_1 = \alpha_1$; put $x_1 = -\lambda_1^{-1}\log(1 - \alpha_1(1 - \exp(-\lambda_1 T)))$. Generate $\xi_i = \alpha_i$ from $U(0, 1)$, $i = 2, ..., m-1$. Put

$$x_i = -\lambda_i^{-1}\log\left(1 - \alpha_i(1 - \exp(\lambda_i(T - x_1 - ... - x_{i-1})))\right). \qquad (7.4.1)$$

The vector $\mathbf{x} = (x_1, x_2, ..., x_{m-1})$ is a realization of $\mathbf{X} = (X_1, X_2, ..., X_{m-1})$.

The explanation follows from the **minitheorem** below:

Minitheorem.

If $\xi \sim U(0, 1)$ then $X_1 = -\lambda_1^{-1}\log[1 - \xi(1 - e^{-\lambda_1 T})]$ has the density

$$f_{X_1}(x) = \frac{\lambda_1 e^{-\lambda_1 x}}{1 - e^{-\lambda_1 T}}, \ x \in [0, T].\# \qquad (7.4.2)$$

$f_{X_1}(x)$ is the density function of the CDF of X_1 conditioned by $(X_1 \leq T)$. This CDF equals $P(X_1 \leq t|X_1 \leq T)$, $t \in [0, T]$.

Prove the minitheorem as an exercise.

7.5 How Large is Variance Reduction Comparing to the CMC?

We did not succeed in proving that our algorithm has bounded relative error when the estimated probability $P(T) = P(\tau_1 + ... + \tau_k \leq T) \to 0$.

Nevertheless, it has very good performance characteristics. We are able to prove the following

Claim 7.5.1

If for all $i = 1, 2, ..., k$, $F_i(T) < \alpha < 1$, then

$$\frac{Var[B_k(T)]}{Var[CMC]} \leq \alpha^k. \qquad (7.5.1)$$

Proof.

$$Var[B_k(T)] = E[\prod_{i=1}^{k} F_i^2(\cdot)] - E^2[\prod_{i=1}^{k} F_i(\cdot)] \qquad (7.5.2)$$

$$\leq \alpha^k (E[\prod_{i=1}^{k} F_i(\cdot)] - E^2[\prod_{i=1}^{k} F_i(\cdot)])$$

$$= \alpha^k E[\prod_{i}^{k} F_i(T)(1 - E[\prod_{i}^{k} F_i(T)]) = \alpha^k P(T)(1 - P(T)).$$

.

Since $Var[CMC] = P(T)(1 - P(T))$, the claim follows.#

Example 7.5.1. Let us compare numerically our algorithm with a CMC for a very simple case of finding $A = P(X_1 + X_2 \leq 0.05)$ for the case of $X_i \sim U(0, 1)$, $i = 1, 2$. $Y = X_1 + X_2$ has a so-called triangular distribution with density function $f(t) = t$ for $t \in [0, 1]$ and $f(t) = 2 - t$, for $t \in [1, 2]$. So, we already know the answer: $A = 0.05^2/2 = 0.00125$.

CMC works as follows: generate $X_1 \sim U(0, 1)$, $i = 1, 2$. If $X_1 + X_2 \leq 0.05$ set $Q := Q + 1$. Repeat this N times. Estimate A as $\hat{A} = Q/N$.

Here are the results of 5 experimental runs each of $Q = 1000$ trials:

0.002, 0.001, 0.001, 0.000, 0.002,

with $\hat{A} = 0.0012$. The result is close to the true value but the individual runs have large variability. Sample standard deviation here equals **0.000840**.

Now let us use our algorithm. By (7.3.5) one replica of unbiased estimate of A is $B_2(0.05) = F(0.05)F(0.05 - v)$, where $F(t) = t$ (uniform distribution!), and v is a realization of random variable $V \sim U(0, 0.05)$, see (7.3.3). So, our algorithm works as follows. Generate one replica of V, call it v and calculate one replica of $B_2(0.05) = 0.05 \cdot (0.05 - v)$. Set $SUM := SUM + B_2(0.05)$, repeat this Q times and estimate $\hat{A} = SUM/Q$.

Here are the results of 5 experimental runs each of $Q = 1000$ trials:

$$0.001239, 0.001249, 0.001217, 0.001268, 0.001274,$$

and $\hat{A} = 0.001249$. The result is very close to the true value. Sample standard deviation here equals **0.000024**, almost 35 times smaller than for the CMC! This means variance reduction of about 1200 times. According to (7.5.1) it should be at least 400 times.

7.6 Importance Sampling in Monte Carlo

Let us forget for a while our particular situation with estimating convolutions and have a more general look at Monte Carlo method.

Suppose we want to estimate $A = E[\psi(X)]$, where the r.v. X has a known density $f(v)$. All Monte Carlo applications can be formulated in terms of finding an estimate of mean value for an appropriately chosen random variable. Then first what comes to mind is to generate random replicas of the r.v. X, $X_1, X_2, ..., X_N$, and to consider the following estimator

$$\hat{A} = \frac{\sum_{i=1}^{N} \psi(X_i)}{N}. \tag{7.6.1}$$

Obviously, $\hat{A}$ is an unbiased estimator of A.

A has the following mathematical expression:

$$E[\psi(X)] = \int_{v \in \Omega} \psi(v) f(v) dv. \tag{7.6.2}$$

(In applications, X may be a k-dimensional vector and then Ω is a region in k-dimensional space.)

Here comes a very simple but important trick, see [43], p. 167. Multiply and divide the expression in the integral in (7.6.2) by a density function $g(v) > 0$ of some random variable Y:

$$A = E[\psi(X)] = \int_{v \in \Omega} \psi(v) f(v) dv = \int_{v \in \Omega} \frac{f(v)\psi(v)}{g(v)} \cdot g(v) dv. \tag{7.6.3}$$

We see from the last expression that now A is represented as an expectation of another random variable $Z = [f(Y) \cdot \psi(Y)]/g(Y)$:

$$A = E[\frac{f(Y)\psi(Y)}{g(Y)}], \ Y \sim g(v). \tag{7.6.4}$$

The choice of $g(v)$ is up to us and it could be done to minimize the variability of $\psi(\cdot)f(\cdot)/g(\cdot)$. Intuitively, it is clear that the choice of $g(v)$ must be done in such a way that the value of g must be small when $\psi \cdot f$ is small, and g must be large if $\psi \cdot f$ is large.

In representing $F^{(m)}(T)$ as the mean value of $B_m(T)$, we achieved this goal by introducing new random variables via their joint density, see (7.3). Originally, in the CMC, $F^{(m)}(T)$ is the mean value of the indicator 0-1 random variable I. After introducing new r.v. $B_k(T)$, we see from Claim 7.3.1 that its mean value is an integral over the new density of random vector (7.3.1), and $B_k(T)$ now **is not** a binary 0-1 variable.

7.7   Problems and Exercises

1. Let $X \sim U(0, T)$ and $0 < T_1 < T$. Find the CDF of X. Prove that the CDF of X **given that** $X \le T_1$ coincides with the CDF of $X^\star \sim U(0, T_1)$.

2. Prove the minitheorem in Section 7.4.

3. Prove the following theorem. Let r.v. $V \sim U(0, 1)$, and r.v. Y has CDF $F(x)$, where $F(x)$ is a continuous strictly increasing function. Then the CDF of the random variable

$$X = F^{-1}(V) \tag{7.7.1}$$

is $F(t)$, i.e. $P(X \le t) = F(t)$.

4. Let $Y \sim W(\lambda, \beta)$, i.e. $P(Y \le t) = 1 - e^{-\lambda t^\beta}$. Prove that the r.v. $Z = Y^\beta$ is exponentially distributed.

5. Let $F(t) = 1 - exp^{-\lambda t^\beta}$, $t \ge 0$ be the so-called Weibull distribution function. β is called *shape parameter*. If $\beta = 1$, Weibull distribution becomes exponential distribution. Suggest an algorithm for generating Weibull random variables. Assume that you have a random number generator for $X \sim (0, 1)$.

6. Let $X_i \sim W(1, \beta), i = 1, 2, 3$, and $Y_3 = X_1 + X_2 + X_3$. Develop a "straightforward" algorithm for an unbiased estimation of $P(Y_3 \le T)$.

7. Consider the algorithm for generating the convolutions, see Section 7.4. Investigate its complexity and show that it is $O(r)$ where r is the number of summands in the convolution.

Chapter 8

Network Lifetime in the Process of Its Destruction

Nothing is more valuable as a workable idea.

Quips and Quotes

8.1 Introduction

In this chapter we will consider a nonrenewable network with arbitrary distributed lifetimes of unreliable edges or unreliable nodes. In Section 2, we consider a T-terminal network with reliable nodes and unreliable edges. The main object of our interest will be network lifetime distribution. The lifetime is defined as the time span starting at $t = 0$, when all network components are up, and terminating at the network entrance into its $DOWN$ state. We present two Monte Carlo algorithms for estimating network lifetime, the first one being quite straightforward and primitive, and the second one - its refinement based on an important combinatorial fact (Lomonosov's Lemma). In Section 3 we generalize the approach of Section 2 to the case of unreliable nodes. Section 4 is devoted to a special case of a network with identically distributed edge lifetimes (and reliable nodes). We suggest an efficient Monte Carlo algorithm for estimating an interesting and important combinatorial object - network *destruction spectrum*, so-called D-spectrum. D-spectrum allows obtaining the number of minimal size min-cuts in the network and provides a useful formula for estimating network lifetime in the process of its edge or node destruction. In a similar way, all the theory works for reliable edges and identical unreliable nodes.

The reader will recognize that a part of the material in this chapter is already partially familiar to her/him from Chapter 6.

Let us remind what is a T-terminal network and the definition of its reliability.

A network $\mathbf{N}$ is an undirected graph $G = (V, E)$ with node set V, $|V| = n$, edge set E, $|E| = m$, and a distinguished set of nodes T, $T \subseteq V$, called *terminals*.

Suppose first that nodes never fail while edges are subject to failures.

The state of $\mathbf{N}$ is defined via the set $X \subseteq E$ as being UP if any two terminals are connected by a path consisting of edges from the set X. If edges from X do not provide connection between all terminals, this state is called $DOWN$.

At the instant $t = 0$, all edges exist. Let edge e be erased (fails) at some random instant $\tau(e) \sim F_e(t)$. Edges fail independently of each other. Edges are nonrenewable, i.e. after edge fails, it remains erased "forever".

Obviously, at $t = 0$ the network is UP. Network lifetime $\tau^\star$ is defined as the random instant at which the network enters the $DOWN$ state. More formally, on $[0, \tau^\star)$ the network is UP, and on $[\tau^\star, \infty)$ the network is $DOWN$.

8.2 Estimation of $F_{\mathbf{N}}(t) = P(\tau^\star \leq t)$

Straightforward estimation of network lifetime distribution
$$F_{\mathbf{N}}(t) = P(\tau^\star \leq t)$$
can be carried out by the following algorithm.

Algorithm 8.1 - Netlife.
1. **For each** $e \in E$, **simulate** edge e lifetime t_e
   according to its CDF $F_e(t)$.
2. **Order** the edge lifetimes in ascending order
   (edges are numbered $1, 2, ..., m$):
   $t_{i_1} \leq t_{i_2} \leq t_{i_3} \leq \ldots \leq t_{i_m}$.
3. **Check** the T-terminal connectivity at each of the instants t_{i_k},
   $k = 1, ...m$, and **determine** the network lifetime,
   i.e. the instant $t^\star$ when $\mathbf{N}$ enters $DOWN$.
4. **Repeat** steps 1,2,3 M times.
5. **Order** M replicas of $t^\star$, $t_1^\star, ..., t_M^\star$ in ascending order.

6. Estimate $\hat{F}_{\mathbf{N}}(t)$ as follows:

$$\hat{F}_{\mathbf{N}}(t) = \frac{\#\text{of } t_i^{\star} \leq t}{M}, \qquad\qquad t = \Delta, 2\Delta, ..., r\Delta, \qquad (8.2.1)$$

where Δ is some fixed positive number.

7. Estimate the variance of $\hat{F}_{\mathbf{N}}(t)$ as

$$\hat{V}ar[\hat{F}_{\mathbf{N}}(t)] = \frac{\hat{F}_{\mathbf{N}}(t)(1 - \hat{F}_{\mathbf{N}}(t))}{M}. \qquad\qquad (8.2.2)$$

Crucial improvement of Algorithm 8.1 - Netlife is achieved by a modification of Step 3. A considerable time saving is provided by applying an important combinatorial fact first discovered by M. Lomonosov [12, 23].

Now we need to repeat some material presented in Chapter 2 regarding trees.

We will introduce some definitions. For each connected graph $G(V, E)$ it is possible to define the so-called *spanning tree* $ST(G)$: this is a *tree*, i.e. a connected graph with no cycles, which connects all vertices of G. Suppose that for each edge $e \in E$ there is defined a nonnegative number $w(e)$ called the *weight* of e. Then the weight $W(ST)$ is defined as the sum of its edge weights:

$$W(ST) = \sum_{e \in ST} w(e). \qquad\qquad (8.2.3)$$

The spanning tree with maximal (minimal) weight is called the maximal (minimal) spanning tree. The minimal spanning tree we denoted as *MST* and the maximal as *MaxST*.

The *terminal* spanning tree is a tree $ST(T)$ that connects all terminals and has no redundant edges in the following sense: elimination of any of its edges disrupts the connectivity of the terminal set T. For example, Fig. 8.1 shows a network, its maximal spanning tree, and its subtree which connects the terminals.

The *MaxST* is constructed by the famous Kruskal algorithm [34], see Chapter 2. On each step of its work this algorithm joins to the tree "under construction" one edge, in the descending order of their weights, if and only if this edge does not create a cycle together with the edges already belonging to the constructed tree.

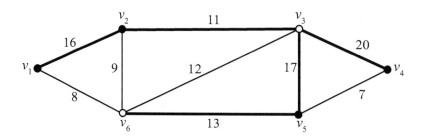

Figure 8.1: $T = \{v_1, v_2, v_4, v_5\}$,
$Max^{(min)}ST(T) = \{(v_1, v_2), (v_2, v_3), (v_3, v_4), (v_3, v_5)\}$

Suppose that network edge e has lifetime $t(e)$. Suppose that all $t(e), e \in E$ are distinct. Assign the weight $w(e) = t(e)$ and consider the maximal-weight spanning tree $MaxST(G)$. Let $Max^{(min)}ST(T)$ be its minimal subset which spans all terminals in T.

Lemma 8.2.1 (Lomonosov [12, 23]).
The network lifetime equals to the smallest lifetime of the edges constituting $Max^{(min)}ST(T).\#$

Fig. 8.1 illustrates this lemma. The minimal subtree $Max^{(min)}ST(T)$ is obtained from the $MaxST(G)$ by eliminating "hanging" edge $e = \{v_6, v_5\}$. The shortest lifetime in the $Max^{(min)}ST(T)$ equals $t^* = 11$ and exactly at this instant the network loses its terminal connectivity.

Let us sketch the proof of the Lemma. Consider $MaxST(G)$. Let its minimal edge lifetime be $t(e^*)$. Obviously network lifetime is at least equal $t(e^*)$. Suppose that the network is alive at time $\theta > t(e^*)$. Eliminate edge e^*. Let this edge connect nodes a and b. The node-set V then "falls apart" into two connected components $V(a)$ and $V(b)$, one containing node a and another containing node b. If the network is still alive at $t(e^*)$, then there must be an edge connecting $V(a)$ and $V(b)$, with weight θ greater than $t(e^*)$. There exists therefore another spanning tree, whose weight is greater than the existing one - contradiction.

The importance of the above lemma lies in the fact that in order to determine network lifetime t^* (Step 3 of Algorithm 1), it is necessary to construct *only one* spanning tree.

Algorithm 8.2 - NetlifeModified [23].
1. **For each** $e \in E$, **simulate** edge e lifetime t_e according

to its c.d.f. $F_e(t)$.

2. **Set** $w(e) := t(e), e \in E$. **Construct**, using Kruskal algorithm, the maximal-weight spanning tree $MaxST(G)$.

3. **Eliminate** hanging edges and **obtain** $Max^{(min)}ST(T)$, the minimal subtree of $MaxST(G)$, which connects all terminals.

4. **Find** the minimal-weight edge in $Max^{(mim)}ST(T)$. **Denote** its weight by $\tau^\star$.

5. **Repeat** steps 1-4 M times.

6. **Order** M replicas of network lifetime:
$$\tau_{i_1}^\star < \tau_{i_2}^\star < \ldots \tau_{i_M}^\star$$

7. **Estimate** $\hat{F}_N(t)$ as follows

$$\hat{F}_N(t) = \frac{\#of \ \tau_{i_j}^\star \leq t}{M}, \qquad t = \Delta, 2\Delta, ..., r\Delta. \qquad (8.2.4)$$

8. **Estimate** the variance of $\hat{F}_N(t)$ as

$$\hat{V}ar[\hat{F}_N(t)] = \frac{\hat{F}_N(t)(1 - \hat{F}_N(t))}{M}. \qquad (8.2.5)$$

Let us compare the efficiencies of Algorithms 8.1 and 8.2. Algorithm 8.1 is a typical example of a straightforward approach to simulation, and a comparison of its efficiency with Algorithm 8.2 will show the gain obtained using Lomonosov's lemma. For Algorithm 8.1, Step 1 needs $O(m)$ operations. Step 2 typically needs $O(m \cdot \log n)$ operations. Step 3 implies that we check the T-terminal connectivity $O(m)$ times; each such check demands finding a spanning tree for the terminals (if such does exist). Typically, it consumes $O(m \cdot \log n)$ operations. Therefore the total number of elementary operations for Algorithm 8.1 is

$$O(m) + O(m \cdot \log n) + O(m)O(m \cdot \log n) = O(m^2 \cdot \log n). \qquad (8.2.6)$$

Contrary to this, Step 3 of Algorithm 8.2 is carried out only *once*, and thus the number of elementary operations is

$$O(m) + O(m \cdot \log n) + O(m \cdot \log n) = O(m \cdot \log n). \qquad (8.2.7)$$

A factor of m is, roughly speaking, the gain provided by our combinatorial approach. For an average size dense network with n about 50 and m about 100, the factor of m is a substantial gain!

8.3   Unreliable Nodes

Assume that in addition to failing edges, also the nodes are subject to failures. To be more accurate, we assume that all nodes, except the terminals, have a random life, i.e. node v fails at the instant $\tau_v \sim F_v(t)$, the nodes fail independently. Besides, edge and node failures are independent events.

Let v be a fixed node, not a terminal; the edges $E(v) = [e_1(v), ..., e_k(v)]$ are incident to v. By our definition, the failure of node v *is equivalent to the failure of all edges $E(v)$ incident to v.* In terms of computer networks, nodes typically represent computers, computer centers, information retransmission devices, etc., while the edges represent information transmission channels. We assume, therefore, that node v failure or the failure of all edges $E(v)$ have the same consequences. In other words, we can assume that the nodes remain intact, but at the instant τ_v all edges $E(v)$ incident to v fail.

If edges *and* nodes fail, there remains one complication. Any edge e connects two nodes, $a(e)$ and $b(e)$. If nodes are subject to failure, then there are three events which might determine edge e lifetime: the failures of nodes $a(e)$ and $b(e)$ and the failure of the edge e itself. Thus the **first** of these events will be the actual moment when edge e goes down. In other words, we can assume that the nodes do not fail but the "true" instant of edge e failure $\theta(e)$ is defined as

$$\theta(e) = \min[\tau_e, \tau(a(e)), \tau(b(e))]. \tag{8.3.1}$$

To treat the case of failing nodes and edges, the changes in Algorithm 8.2 are obvious. Step 1 is replaced by Steps 1a, 1b, and 1c:

1a. For each $e \in E$, **simulate** lifetime $\tau_e = t_e$.
1b. For each nonterminal node v, **simulate** its lifetime $\tau_v = t_v$.
1c. For each $e \in E$, **recalculate** its lifetime as
the minimum of $t_e, t(a(e)), t(b(e))$.

The number of elementary operations in the modified Algorithm 8.2 is $O(m \cdot \log n)$, i.e. of the same order as in the original version.

The following remark establishes an important connection between the destruction process and the static situation.

Remark: network static reliability. Suppose that network edges fail independently, and let us consider a **static** situation in which edge $e \in E$ fails with probability $q_e = 1 - p_e$. Imagine that in our dynamic destruction process we take for edge e lifetime such CDF $F_e(t)$ that $F_e(t_0) = q_e$ for all $e \in E$. Now $F_{\mathbf{N}}(t_0)$ is, on one hand, the probability that the network is *DOWN* at $t = t_0$ in our destruction process. On the other hand, at $t = t_0$ each edge is *down* with probability which coincides with its static down probability. Therefore, $F_\mathbf{N}(t_0)$ **equals network static** $DOWN$ **probability.**#

8.4 Identically Distributed Edge Lifetimes

In this section we return to the network with reliable nodes and add one important assumption: all edges have the same lifetime $\tau_e \sim F_0(t)$. (Edge failures remain independent events.) The material below will be familiar to the reader from Chapter 6.

Consider edge network "destruction" process as it evolves in time. Let edge failures appear at the instants θ_i:

$$\theta_1 < \theta_2 < \ldots < \theta_m. \tag{8.4.1}$$

Let θ_k be the first instant at which the network fails. In other words, the network is UP on $[0, \theta_k)$ and $DOWN$ on $[\theta_k, \infty)$. What interests us is not the value θ_k but the **index** at the θ. This index k is **random**, because it depends on the random order in which the edges fail. To clarify this important point, let us consider a four edge network shown on Fig. 8.2 (this example is identical to the example shown on Fig. 6.3).

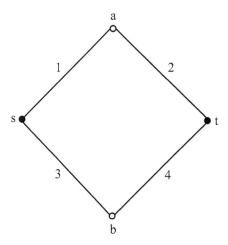

Figure 8.2: $s - t$ connection fails at the instant of the second or third edge failure, depending on the order in which edge failures take place

The terminals are s and t. There are $24 = 4!$ different equally probable orders in which edges fail. Direct enumeration shows that 16 of them lead to the network failure at the instant of the second edge failure, e.g. if the order of edge failures is e_1, e_3, e_2, e_4. Thus $k = 2$ with probability 16/24=2/3. s and t are always disconnected after the third failure. Thus $P(k = 3) = 8/24 = 1/3$.

Randomness of the number k at which the network fails is created by the random order of edge failure appearance. Since all edge lifetimes are identical and failures appear independently of each other, all $m!$ such orders are *equally probable*. The following definition is already familiar to the reader from Chapter 6 and is identical to Definition 6.3.1.

Definition 8.4.1

The distribution of k, i.e. the collection of probabilities f_j

$$S_p = [f_1, f_2, ..., f_m]$$

is called *D-spectrum* of the network.#

S_p was termed in [11] *internal distribution* of the network.

The D-spectrum depends on $G = (V, E)$ and the terminal set T. So, for example, if the terminal set for the network on Fig. 8.2 is $T = \{a, b, s, t\}$, then the loss of connectivity takes places always on the second edge failure. Therefore, the D-spectrum consists of one point $f_2 = 1$.

A trivial but very important fact is that, by the above definition, the D-spectrum does **not depend** on the edge lifetime distribution $F_0(t)$.

Consider any particular order of edge failure appearance, e.g. $i_1, i_2, \ldots, i_m$, and consider a set of order statistics from the population with CDF $F_0(t)$

$$\tau_{(1)} \leq \tau_{(2)} \leq \tau_{(3)} \leq \cdots \tau_{(m)}. \tag{8.4.2}$$

The moment at which the first edge i_1 fails is $\tau_{(1)}$; the moment at which edge i_2 fails is $\tau_{(2)}$, etc. The moment at which the r-th edge fails coincides with $\tau_{(r)}$, $r = 1, ..., m$. Note that $\tau_{(r)}$ has the CDF given by the following formula (see Chapter 6):

$$F_{(r)}(t) = \sum_{j=r}^{m} \frac{m!}{j!(m-j!)} [F_0(t)]^j [(1 - F_0(t))]^{m-j}. \tag{8.4.3}$$

The central result of this section is identical to Claim 6.1.1.

Claim 8.4.1

For identical edge lifetimes, network lifetime $\tau_{\mathbf{N}}$ has the following distribution:

$$P(\tau_{\mathbf{N}} \leq t) = F_{\mathbf{N}}(t) = \sum_{r=1}^{m} f_r F_{(r)}(t). \tag{8.4.4}$$

The proof follows immediately from the definitions of the spectrum and the order statistics, by using the total probability formula.#

The importance of Claim 8.4.1 is determined by the fact that in order to estimate network reliability for any $F_0(t)$ and any t, we need to estimate only network spectrum and later act according to (8.4.4).

Algorithm 8.3 - NetlifeSpectrum.
1. **Set** $N_1 = N_2 = ... = N_m = 0$.
2. **Simulate** random permutation π for the numbers
   $1, 2, ..., m$, $\pi = (i_1, i_2, ..., i_m)$.
3. **Assign** weight r, to the r-th edge i_r, $r = 1, ..., m$.
4. Using Kruskal's algorithm, **find** the maximal-weight
   spanning tree for the network.
5. **Eliminate** hanging nodes and **find out** the minimal
   subtree $Max^{(min)}ST(T)$ of the terminal set.
6. **Find** the minimal-weight edge in $Max^{(min)}ST(T)$. Let its weight
   be $w^\star$, $1 \le w^\star \le m$.
7. **Let** $q := w^\star$; **Put** $N_q := N_q + 1$.
8. **Repeat** Steps 2-7 M times.
9. **For** $r = 1, ..., m$, **put** $\hat{f}_r = N_r/M$.
10. **Estimate** $F_{\mathbf{N}}(t)$ as

$$\hat{F}_{\mathbf{N}}(t) = \sum_{r=1}^{m} \hat{f}_r F_{(r)}(t). \tag{8.4.5}$$

This algorithm checks the terminal connectivity in the edge "destruction process" according to the order given by π. It is separated from the process of edge failures in real time, and thus the D-spectrum becomes a combinatorial characteristic of the network. $\hat{f}_r$ estimates the probability of network failure after erasing the r-th edge i_r in π. The algorithmic trick in NetlifeSpectrum is finding the "critical" edge by constructing a single ST - see Step 4.

Representation of $\hat{F}_{\mathbf{N}}(t)$ via the estimated spectrum

$$\hat{S}_p = [\hat{f}_1, \hat{f}_2, ..., \hat{f}_m]$$

is very convenient for variance estimation. We have the following

Claim 8.4.2

$$Var[\hat{F}_{\mathbf{N}}(t)] = \qquad (8.4.6)$$

$$\Big(\sum_{r=1}^{m} f_r(1-f_r)F_{(r)}^2(t) - 2\sum_{i<j} F_{(i)}(t)F_{(j)}(t)f_if_j \Big)M^{-1}.$$

We omit the proof. It is based on the properties of covariances for multinomial distribution, see Section 1.4.#

Since the second term in (8.4.6) is negative, the variance does not exceed the first sum in (8.4.6). Then it is easy to obtain an upper bound on the relative error

$$r.e. = \frac{Var[\hat{F}_{\mathbf{N}}(t)])^{0.5}}{E[\hat{F}_{\mathbf{N}}(t)]} \le (\max_k f_k)^{0.5}/\min_k f_k \le Const. \qquad (8.4.7)$$

We see from this expression that $F_{(r)}(t)$ are not presented in the bound on the *r.e.*. So, it remains bounded if $t \to 0$, or $t \to \infty$, i.e. for very small and very large edge failure probabilities.

Remark 1. Suppose, we can not assume that all edges (nodes) have *equal* failure probability. The only thing we know is that *all* $p_i \in [p_{\min}, p_{\max}]$. Then, obviously, the network failure probability $F_{\mathbf{N}}$ lies in the interval $[F_{\mathbf{N}}(\min), F_{\mathbf{N}}(\max)]$, where $F_{\mathbf{N}}(\min)$ is obtained from (8.4.5) by setting all $p_i = p_{\max}$ and $F_{\mathbf{N}}(\max)$ is obtained from (8.4.5) by setting $p_i = p_{\min}$.#

Remark 2: Reliable edges, unreliable nodes. Suppose we assume that edges do not fail, some nodes which are terminals also do not fail, but the remaining nodes have i.i.d. lifetimes. If a node fails, all edges incident to it are erased. (We exclude the situation when there are only two terminals directly connected by an edge so that this network never fails.) At $t = 0$ all nodes are up, at some instants $\theta_1, \theta_2, ...$, nodes fail and at some instant $\tau^\star$ the network becomes *DOWN*. All the theory developed earlier for unreliable edges remains valid for the case of unreliable nodes. Similar to the edge D-spectrum, we will find the node D-spectrum. In Chapters 10 and 13 we will present examples of node D-spectra and consider examples of computer networks, in which the nodes and not the edges are subject to failure.#

8.5 Use of D-Spectra to Estimate the Number of Min Size Min-cuts

Figures 8.3(a), 8.3(b), 8.3(c), 8.3(d) present four small networks together with their edge D-spectra.

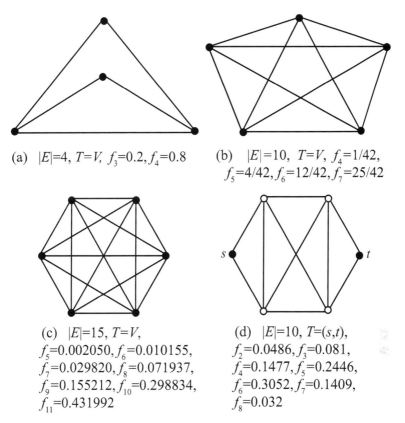

(a) $|E|=4$, $T=V$, $f_3=0.2, f_4=0.8$

(b) $|E|=10$, $T=V$, $f_4=1/42$, $f_5=4/42, f_6=12/42, f_7=25/42$

(c) $|E|=15$, $T=V$, $f_5=0.002050, f_6=0.010155$, $f_7=0.029820, f_8=0.071937$, $f_9=0.155212, f_{10}=0.298834$, $f_{11}=0.431992$

(d) $|E|=10$, $T=(s,t)$, $f_2=0.0486, f_3=0.081$, $f_4=0.1477, f_5=0.2446$, $f_6=0.3052, f_7=0.1409$, $f_8=0.032$

Figure 8.3: Examples of edge D-spectra

Case 8.3(a) is a four node, five edge network, with terminal set $T = V$. Case 8.3(b) is a complete five node graph, all nodes are terminals.

8.3(c) is a six node complete graph, with $T = V$. 8.3(d) presents a ten edge network, $T = (s,t)$. Spectra were estimated using $M = 10,000$ replications.

Let us show how the D-spectrum can be used to estimate the number of minimal size min-cuts W in the network. We remind that we already discussed this problem in Chapter 6, Section 4.

The formula we need is the following:

$$W = f_r \frac{m!}{r!(m-r)!},$$ (8.5.1)

where f_r is the first *nonzero* element of the spectrum.

Let us consider the network on Fig. 8.3(c). The first nonzero term in the spectrum is $f_5 = 0.00205$, $m = 15$, $r = 5$. By (8.5.1) we obtain $W = 6.15$ which we round to 6. And this is exactly the number of minimal cuts of minimal size 6 in the network! (These cuts isolate each of 6 nodes by deleting the edges adjacent to these nodes.)

Table 2 of paper [11] presents the D-spectrum of the dodecahedron, see Fig. 4.2, for two cases: a) $T = V$ and b) $T = \{1, 7, 8, 11, 16\}$, see also Appendix C.

For case a), the first nonzero term in the spectrum is $\hat{f}_3 = 0.00476$ obtained from simulating $M = 10^5$ permutations. By (8.5.1) the number of min-cuts of minimal size (three) is $\hat{W} = 0.0047630!/(3! \cdot 27!) = 19.3$. The correct number of minimal size min-cuts is 20. These min-cuts isolate each of the nodes from others.

For case b), $\hat{F}_3 = 0.00130$, the true number of min size min-cuts is five. (8.5.1) gives 5.3.

Let us consider next several examples of estimating the number of min size cuts in relatively large networks. Suppose we investigate a network with $m = 80$ edges and construct the spectrum on the basis of doing M experiments, i.e. we generate M permutations. Suppose we found that the minimal size of a cut set is $r = 5$, i.e. the first nonzero element of the spectrum is f_5. Let us estimate the probability that we *missed* a minimal cut set of size $r = 4$.

The probability to reveal a cut set of size 4 in a single experiment by generating a random permutation of size 80 is

$$\varepsilon = \frac{4! \cdot 76!}{80!} = 6.3 \cdot 10^{-7}$$ (8.5.2)

Therefore to *miss* the cut set of size 4 in $M = 10^7$ independent experiments is

$$P_{miss} = (1 - \varepsilon)^M \approx e^{-6.3} = 0.0018.$$ (8.5.3)

The first experiment is a hypercube H_5 which has 32 nodes and 80 edges. (A hypercube H_k has 2^k nodes and $k \cdot 2^{k-1}$ edges.) Edges are subject to failure, nodes do not fail. The network has 12 randomly chosen nodes as the terminals, and the reliability criteria is the terminal connectivity. We have carried out 10 simulation runs, each run of 10,000,000 replicas.

It was found that the minimal cut dimension is $r = 5$. Since each run gave a different estimate of the number of min-cut sets, we constructed a 95 % Student small-sample confidence interval of the mean number of cuts $W(5)$, based on the results obtained in the 10 runs. This confidence interval was [11.95, 12.75] and we assumed $W(5) = 12$. For this number of min-cuts, we computed the Burtin-Pittel approximation to the network reliability (denoted as $B - P$), together with the estimate of network failure probability F by using (8.4.4) and the estimated network spectrum. Below are the numerical results for various edge failure probabilities $F(t) = \alpha$:

For $\alpha = 0.4$, $\hat{F} = 0.137$, $B - P = 0.123$;

For $\alpha = 0.2$, $\hat{F}_N = 0.00394$, $B - P = 0.00384$;

For $\alpha = 0.1$, $\hat{F}_N = 0.000121$, $B - P = 0.00012$;

For $\alpha = 0.05$, $\hat{F}_N = 3.8 \cdot 10^{-6}$, $B - P = 3.7 \cdot 10^{-6}$.

In this example the $B - P$ approximation provides an extremely accurate estimate of the network failure probability.

Our next experiment was with the same H_5 network, 12 terminals and 80 edges, 30 of which being randomly displaced. The minimal r was found to be $r = 3$, in a run consisting of 100,000,000 replications. Only a *single* minimal min size cut of $r = 3$ was in this run.

Below are the results of computing an estimate for the exact network failure probability via the spectrum and the $B-P$ approximation, for various α values:

For $\alpha = 0.2$, $\hat{F}_N = 0.0151$, $B - P = 0.008$;

For $\alpha = 0.1$, $\hat{F}_N = 0.00137$, $B - P = 0.001$;

For $\alpha = 0.05$, $\hat{F}_N = 0.000146$, $B - P = 0.000125$;

For $\alpha = 0.01$, $\hat{F}_N = 1.05 \cdot 10^{-6}$, $B - P = 1.0 \cdot 10^{-6}$.

Here the B-P approximation is quite satisfactory for small $\alpha < 0.1$.

(8.4.6) allows estimating the accuracy of network reliability using the estimated spectrum. For a run of 10,000,000 permutations, for the H_5 network, and edge failure probability $\alpha = 0.2$, we have found that $\hat{F}_N = 0.00393$, and the estimate of the corresponding standard deviation $\hat{\sigma} = 0.000013$. Similarly, when $\hat{F}_N = 0.000121$, we found $\hat{\sigma} = 0.0000016$. The above data allow to conclude that the network reliability estimates are quite accurate.

The estimate of system D-spectrum allows estimating the number $C(x)$ of system down states with exactly x down components. This fact is formulated in a form of the following

Claim 8.5.1.

Let $S_p = [f_1, f_2, ..., f_m]$ be the D-spectrum of a monotone system consisting of m components. Denote by $C(x)$ the number of its down states with exactly x components being down and $m - x$ being up. Then

$$C(x) = (f_1 + f_2 + ... + f_x)\frac{m!}{x! \cdot (m - x)!}\#.$$ (8.5.4)

The idea of the proof is simple. Suppose that all components fail independently and have equal failure probability α. All failure states may be classified into different classes according to the number of down elements in the failure state. In this way we obtain system failure probability in the following form (compare with (4.4.5)):

$$F = \sum_{k=1}^{x} C(k)\alpha^k(1 - \alpha)^{m-k}.$$ (8.5.5)

On the other hand, we have the spectral representation (8.4.4) of the failure probability (with $\alpha = F_0(t)$, see (8.4.3)). We will obtain the desired representation (8.5.4) by comparing the coefficients at $\alpha^x(1 - \alpha)^{m-x}$ in both representations.

The formula (8.5.4) establishes a combinatorial fact. Let us present an alternative proof of it using only combinatorial arguments.

Consider a random permutation $\pi = (i_1, i_2, ..., i_m)$. Declare the first x of its members as system's component numbers which are *down* and all the rest as being *up*. If this permutation determines system $DOWN$ state, call it $(x; D)$-type permutation. Denote by $N(x)$ the total number of $(x; D)$ permutations. Obviously, the probability to have an $(x; D)$-type permutation is $N(x)/m!$.

On the other hand, this probability equals $f_1 + f_2 + ... + f_x$. It follows from the definition of the destruction process. Suppose, a permutation has the property that the system failure was observed at the instant of k-th failure, $1 \le k \le x$. Declare for this permutation all components whose numbers appear on the next $x - k$ positions as being *down*, and all the rest of components as being *up*. In this way we will reconstruct all permutations of $(x; D)$-type. Note also that any permutation which in the destruction process produces $DOWN$ *after* the x-th step is *not* of $(x; D)$-type. Therefore, $N(x) = (f_1 + f_2 + ... + f - x)m!$.

When we define a system $DOWN$ state with exactly x components being *down*, the order of their appearance is not relevant. All permutations obtained by permuting x *down* components between themselves and $(m - x)$ between themselves determine, in fact, the same system failure state. Therefore, $C(x) = N(x)/(x!(m - x)!).\#$

Remark. Imagine a system of m independent components which all are *up* at $t = 0$. Each component, independently of others, goes *down* (and remains forever *down*) at some random instant τ_i in such a way that its probability to be *down* at t_0 equals $q_i = P(\tau_i \leq t_0)$, $i = 1, 2, ..., m$. As the result of this destruction process, at t_0 the system will be either UP or $DOWN$. Let $P(\text{system is } DOWN \text{ at } t_0) = Q_d(t_0)$. If τ_i-s are i.i.d., then all $q_i = q$.

Now imagine a binary lottery for each of system components. Component i, $i = 1, 2, ..., m$, *independently of others*, is declared *down* with probability q_i and *up* with probability $p_i = 1 - q_i$. The result of these lotteries will be UP or $DOWN$ for the whole system. Let $P(\text{lotteries produce } DOWN) = Q_L$. Q_L is system *static* failure probability.

The following fact is the key to computing system reliability using the D-spectrum:

$$Q_d(t_0) = Q_L.$$

Let us go one step further and consider a network with *renewable* components in *stationary* regime. Suppose that each component, independently of others, undergoes an alternating sequence of *up* and *down* states, in such a way that its stationary probability to be *up* equals p, and to be *down* equals $q = 1 - p$, see Section 4.3. All components have, therefore, the same *up* probability. Then the stationary probability that the whole network will be $DOWN$ equals the network *static DOWN* probability. It is easy to conclude that if we set in (8.5.5) $\alpha = q$, then F will express the network stationary $DOWN$ probability, and $P(\text{network is } UP) = 1 - F.\#$

8.6 Problems and Exercises

1. Consider the network shown on Fig. 8.4.

   Edge lifetimes are shown on the figure.

   a). Find the network lifetime, assuming all nodes are absolutely reliable.

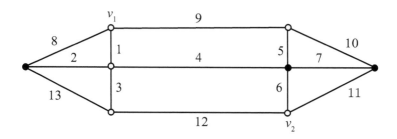

Figure 8.4: Bold nodes are terminals

b). Suppose that in addition to edge failures, node v_1 fails at the instant $x = 6.5$, and node v_2 fails at the instant $y = 10.5$. Find the network lifetime.

(If in the process of constructing the maximal spanning tree you have to choose between two or more edges with equal weights, take arbitrarily any one of them.)

2. Prove (8.4.7).

3. Prove in detail the Claim 8.5.1. Compute the number of *down* states with $x = 7$ components for the network on Fig. 8.3(d).

4. Which number of replications M would provide the probability $p > 0.999$ of not missing a min-size cut of size $r = 3$ in a network with $m = 90$ unreliable edges?

5. Terminal s is connected to node 1, node 1 is connected to terminal t; in parallel, s is connected to node 2, node 2 to node 3, node 3 to node 4, and node 4 is connected to t. The connections do not fail, the nodes are subject to failures. Nodes fail independently and have equal failure probability F. The network fails if the $s - t$ connection is broken.
a. Find the D-spectrum of the network.
b. Find using (8.4.4), the network failure probability $F_{\mathbf{N}}$. Find numerically the maximal permissible value of F which guarantees that F_G does not exceed 0.05.

6. Design an algorithm for generating a random permutation of $n > 3$ integers $(1, 2, 3, ..., n)$. Use the following recursive procedure. Suppose you have a random permutation of k integers $(i_1, i_2, ..., i_k)$. Put the integer $k+1$ at the end of this permutation and exchange it with a randomly chosen number among the first k numbers. Proceed until $k = n$.

7. Consider the five node network on Fig. 8.3(b). Find the number of min-cuts (of size $r = 4$) and the $B - P$ approximation to network failure probability for edge failure probability ϵ. Assuming that the approximation is accurate, find the maximal ϵ which guarantees network failure probability less or equal to 0.01.

8. Write a computer program which calculates system failure probability by (8.4.4) for a given D-spectrum and given component failure probability $F_0(t) = q$.

Chapter 9

Combinatorial and Probabilistic Properties of Lomonosov's "Turnip"

> The real crown jewels are ideas.
>
> Quips and Quotes.

9.1 Introduction

This chapter is devoted to a very powerful and efficient Monte Carlo algorithm for network reliability estimation which we call "Lomonosov's turnip".

The idea of this algorithm was first suggested by M.V. Lomonosov in [36] and developed later in a series of works [11,12,37,38]. The "turnip", called in [36] *evolution process with closure,* was primarily designed for estimating network terminal reliability for the case of arbitrary (and nonequal) network edge failure probabilities, in the static setting. Lomonosov's algorithm introduces closure operation on network edges which eliminates so-called non relevant edges. This allows to accelerate considerably the Monte Carlo simulation process and makes it possible to handle relatively large networks with hundreds of edges. Lomonosov's algorithm uses specially designed *trajectories* leading from the initial "zero" state to network *UP* state. These trajectories allow to identify the "pre-failure" or "border" states of the network and open a way for simulating dynamic network stationary mean *UP* and *DOWN* periods as well as the reliability gradient function.

119

The most important and common network reliability problem is the calculation of its *static reliability*, i.e. the probability that the network is in the UP state. The word "static" means that the time coordinate is not presented here and that each edge, independently of other edges, is subject to a random choice of its state. This state is *up* with probability $1 - q(e) = p(e)$ and *down* with probability $q(e)$. As usual, *down* for an edge means that it is erased. Note that the straightforward calculation of static reliability is very time consuming, even for moderate size networks. For example, the well known path-set or cut-set methods demand computation time which grows exponentially with the size of the network. Computational difficulties in the direct calculation of static network reliability stimulated great interest in developing Monte Carlo methods.

We do not intend to review various approaches in this direction. Interested readers can look into the fundamental monograph [16] and read the paper [11]. In general, most works in this direction are related to the crude Monte Carlo (CMC) and its modifications. These methods have a principal drawback: they are inefficient for large and highly reliable or highly unreliable networks (because of the so-called rare event phenomenon). For most applications, the greatest interest for computer and/or communication networks are highly reliable networks. Let us mention that the Lomonosov's approach eliminates *in principle* the rare event phenomenon.

9.2 The Turnip

9.2.1 The idea of the turnip

The construction which we call "turnip" is based on three ideas. The first is introducing an artificial random process associated with each edge. The second is defining the trajectories of a random process built on network states. The third is using a special combinatorial operation - closure, which considerably accelerates the computation procedure.

9.2.2 Artificial creation process

Associate with each edge an *artificial creation process*, see Fig. 9.1. Initially, at $t = 0$ every edge e is *down*. At some random moment, edge e is "born", independently of others, and remains *up* forever. The random moment $\xi(e)$ of edge "birth" is exponentially distributed:

$$P(\xi(e) \leq t) = 1 - \exp(-\lambda(e) \cdot t), \ e \in E. \tag{9.2.1}$$

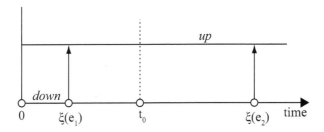

Figure 9.1: At instant t_0, edge e_1 does exist, while edge e_2 does not

Fix an arbitrary moment t_0, in particular let $t_0 = 1$. Choose for each edge e its "birth rate" $\lambda(e)$ so that the following condition holds:

$$P(\xi(e) > t_0) = \exp(-\lambda(e) \cdot t_0) = q(e). \tag{9.2.2}$$

This formula means the following: in the *dynamic* process of edge birth the probability of being *down* at t_0 for each edge **coincides** with the static *down* probability $q(e)$.

At $t = 0$ the network is $DOWN$ (there are no edges). Denote by $\xi(\mathbf{N})$ network's "birth" time, i.e. at random instant $\xi(\mathbf{N})$ when the network becomes UP (and remains there forever). $P(\xi(\mathbf{N}) \leq t_0)$ is the probability that at moment t_0 in the "creation" process the network $\mathbf{N}$ is UP:

$$R(\mathbf{N}) = P(\xi(\mathbf{N}) \leq t_0). \tag{9.2.3}$$

In words: the static probability $R(\mathbf{N})$ that the network is UP coincides with the probability that in the edge birth ("creation") process network state is UP at the instant t_0, i.e. $\mathbf{N}$ became UP before or at t_0. (The reader probably has noticed that this reasoning is similar to the reasoning in Section 8.2 with regard to edge destruction process.)

9.2.3 The closure

One of the most important advantages of the edge creation approach is the possibility to use combinatorial features of the network, namely the operation of *closure*. We will give an intuitive explanation for closure.

Suppose that the edges in the creation process appear randomly and independently. In Fig. 9.2(a) and 9.2(c) a network with four nodes and five edges is presented. We are going to use this network in the following examples, for two operational criteria: all-terminal connectivity and $s - t$ connectivity.

Suppose that edges 1 and 2 are already born. Then at this moment the nodes associated with edge 3 are already connected by a *path* formed by edges 1 and 2. Therefore, what happens to edge 3 (will it be born or not) **does not affect** the connectivity of the component $\{1, 2, 3\}$. Thus edge 3 becomes *irrelevant* for the further evolution of the network and can be ignored. For each state $S \subseteq E$ of the network we define a closure as a subset of edges in E such that it equals to the union of S and all irrelevant edges (for given operational criterion). For example, in Fig. 9.2(b) the set $\sigma_{21} = \{1, 2, 3\}$ is the closure for each of the following states $\{1, 2\}, \{1, 3\}, \{2, 3\}$. Let us carry out the closure for each connected component of the network. (A single node is considered as a closed component.) The collection of all network components after their closure will be called a *super-state*.

9.2.4   Turnip as evolution process with closure

The diagram of the creation (evolution process) is presented on Fig. 9.2(b). The network **N** has four nodes and five edges, see Fig 9.2(a). By definition, its operational criterion is all node connectivity, i.e. it is *UP* if all nodes are connected. Since at the root and on the top of it there is a single super-state, the diagram has a form of a turnip, and from here comes the word "turnip".

Initially (at $t = 0$) all edges are in the down state, i.e. the initial *super-state* is a collection of four one-node components. The zero-level of the turnip (its "root") is the set V without edges (no edge was born). We denote it by σ_0.

The first level shows all possible evolution results arising after the birth of a single edge. There are five such super-states $(\sigma_{11}, ..., \sigma_{15})$ denoted by circles, see Fig. 9.2(b), the first level. No closure can be performed on this level.

The second level of the turnip shows what happens when a second edge is born. Depending on the particular combination of the first two born edges, we distinguish six super-states on level 2: $\sigma_{21}, ..., \sigma_{26}$. It is important to stress that these super-states are shown together with the relevant closure. For example, suppose that after edge 1, edge 3 is born. Then edge 2 becomes irrelevant and is added to the existing nodes, and the corresponding super-state is σ_{21} Similarly, the same σ_{21} appears if edge 2 is born after edge 1.

Note that the zero level represents super-states with four isolated components. Level 1 represents super-states with three isolated components. Level 2 represents all super-states with two components. The last, third level has one super-state containing one single component.

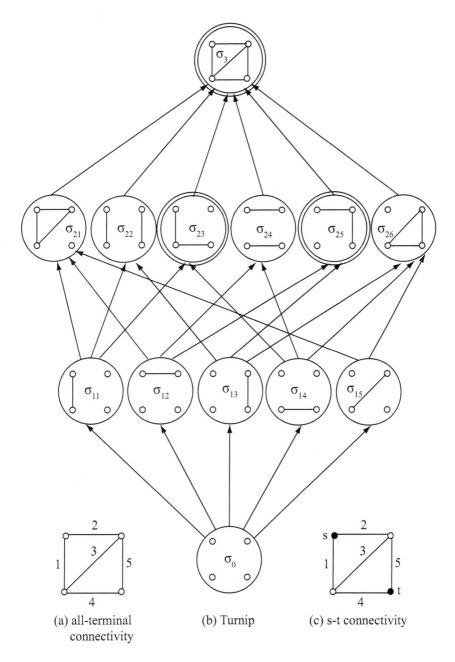

Figure 9.2: The "turnip" diagram for two cases of connectivity

In the case of all-terminal connectivity, only this super-state represents the UP state of the network.

In the case of the $s - t$ connectivity, the UP states are those super-states for which the nodes s, t belong to one component. Such super-states appear on Level 2 and 3. On Fig. 9.2(b) they are double circled.

An important feature of the turnip is that simulating the transitions from one super-state to the following is very easy. Compute, for example the probability $P(\sigma_{11} \rightarrow \sigma_{21})$. Let the birth rate of edge k be $\lambda(k)$, $k = 1, ..., 5$. Then, by the well-known property of the exponential distribution (see Chapter 3), we have

$$P(\sigma_{11} \rightarrow \sigma_{21}) = \frac{\lambda(2) + \lambda(3)}{\lambda(2) + \lambda(3) + \lambda(4) + \lambda(5)} = \frac{\lambda(\sigma_{11}) - \lambda(\sigma_{21})}{\lambda(\sigma_{11})}, \quad (9.2.4)$$

where $\lambda(\sigma_{ij})$ denotes the total birth rate for the super-state σ_{ij}. Indeed, the transition $\sigma_{11} \rightarrow \sigma_{21}$ takes place if and only if either edge 2 was born before the birth of edges 3, 4, 5, or edge 3 was born before the birth of edges 2, 4, 5.

Let us make an important remark. It follows from (9.2.2) that

$$\lambda(e) = -(\log[q(e)])/t_0.$$

Thus t_0 is the same factor appearing in all $\lambda(e)$ and thus the transition probabilities (9.2.4) are t-invariant. Therefore, these probabilities remain the same when t_0 goes to zero or to infinity, i.e. for very reliable or very unreliable edges.

Summing up, the turnip diagram describes an artificial creation process with closure. The probability that the network in this process is UP at some moment t_0 *coincides* with the corresponding static probability for the network.

Consider a random process $\sigma(t)$ whose states are the super-states of the above described network evolution process. (For example, $\sigma(t = 0) = \sigma_0$.) We already mentioned that each state of $\sigma(t)$ is a super-state. The following theorem was proved in [11,12].

Theorem 9.2.1.

(i) $\sigma(t)$ is a Markov process;

(ii) The time spent by $\sigma(t)$ in a particular super-state $\sigma^\star$ is distributed as $Exp(\lambda(\sigma^\star))$;

(iii) Let $\sigma^{\star\star}$ be a direct successor of $\sigma^{\star}$. Then the transition $\sigma^{\star} \to \sigma^{\star\star}$ takes place with probability

$$P(\sigma^{\star} \to \sigma^{\star\star}) = \frac{\lambda(\sigma^{\star}) - \lambda(\sigma^{\star\star})}{\lambda(\sigma^{\star})}. \tag{9.2.5}$$

Remark 1. Each successor $\sigma^{\star\star}$ of $\sigma^{\star}$ is obtained by merging exactly two components of $\sigma^{\star}$. In [36], $\sigma(t)$ was called Merging Process (MP).#

We omit the proof which is very simple since it follows from the properties of the exponential distribution, see Chapter 3.#

Define now a new notion, *trajectory*, which plays the central role in the Merging Process and therefore in the corresponding Monte Carlo scheme.

Definition 9.2.1

A trajectory is a sequence $u = (\sigma_0, \sigma_1, ..., \sigma_r)$ of super-states such that σ_0 is the initial trivial super-state, each σ_i is the direct successor of σ_{i-1}, and σ_r is the first super-state belonging to UP.#

For example, $(\sigma_0, \sigma_{11}, \sigma_{21}, \sigma_3)$ on Fig 9.2(b) is a trajectory.

Now, in terms of the trajectories, the network is UP at moment t if there exists at least one trajectory that reaches UP before t. It is easy to calculate the probability $p(u)$ that the evolution process goes along trajectory u:

$$p(u) = \prod_{i=0}^{r-1} P(\sigma_i \to \sigma_{i+1}). \tag{9.2.6}$$

Denote by $P(t|u)$ the probability that the UP state (i.e. σ_r) will be reached **before** time t **given** that the evolution (the Merging Process) goes along trajectory u. Then by (ii) of Theorem 9.2.1, our MP is sitting in each super-state σ_j an exponentially distributed random time $\tau(\sigma_j)$; due to the Markov property, the total evolution time along the trajectory u is a sum of the respective independent exponentially distributed random variables. More formally,

$$P(t|u) = P[\tau(\sigma_0) + \tau(\sigma_1) + ... + \tau(\sigma_{r-1}) \leq t | u = (\sigma_0, ..., \sigma_r)]. \tag{9.2.7}$$

Note that $P(t|u)$ is a convolution and can be computed analytically (see Appendix B) or simulated (see Chapter 7).

Now we eventually are ready to present the main result of this chapter

Theorem 9.2.2.

The probability that at moment t the MP reaches UP equals, by the total probability formula,

$$R(\mathbf{N}) = P[\xi(\mathbf{N}) \le t] = \sum_{u \in U} p(u) \cdot P(t|u), \qquad (9.2.8)$$

where U is the set of all trajectories.#

The expression (9.2.8) has a form of a mean value since $\sum_{u \in U} p(u) = 1$, and is the key for estimating $R(\mathbf{N})$ by means of the following Monte Carlo algorithm.

Algorithm 9.1 - Algorithm "Turnip"

1. **Put** $\hat{R} := 0$.
2. **Generate** trajectory u leading from the trivial super-state to the super-state in UP. Use for its generation the above described transition probabilities.
3. **Calculate** $\hat{R} := \hat{R} + P(t|u)$.
4. **Repeat** 2 and 3 M times.
5. **Put** $\hat{R}(\mathbf{N}) = \hat{R}/M$.

Remark 2. The trajectory u is drawn with probability (9.2.6) which *does not depend* on parameter t. This explains why in the turnip scheme the rare event phenomenon does not exist.#.

The following theorem was proved in [37,38].

Theorem 9.2.3.

For a given number of nodes and a given operational criterion, the coefficient of variation δ_{MP}^2 is bounded uniformly for all $t \in [0, \infty)$ and all λ-vectors satisfying $\max \lambda / \min \lambda \le C$, for any C.#

Theorem 9.2.3 assures (through the Chebyshev's inequality), that for any given number of nodes n and $\lambda(e)$ values, and for any positive ϵ, δ, there exists sample size M such that for all $t \in [0, \infty)$, we have

$$1 - \epsilon < \frac{\hat{R}(N)}{R(N)} < 1 + \epsilon \qquad (9.2.9)$$

with probability at least $1 - \delta$. Note that this property does not hold for the crude Monte Carlo.

9.3 Applications of Turnip

9.3.1 Availability $Av(\mathbf{N})$

The static reliability $R(\mathbf{N})$ also expresses the equilibrium instant network *UP* probability, termed also *availability* and denoted $Av(\mathbf{N})$. In other words, availability is the probability that the network is *UP* at a remote instant $t \to \infty$.

Suppose that the life of each edge e is described by an alternating renewal process with exponentially distributed *up* and *down* times. Consider the process $E(t)$ defined as the set of all edges in *up* state at moment t, $t > 0$. $E(t)$ alternates between *UP* and *DOWN* states of the network. The stationary availability of the whole network equals $R(\mathbf{N})$ when each edge e has the stationary *down*-probability $q(e) = \lambda^\star(e)/(\lambda^\star(e) + \mu(e))$, where $\lambda^\star(e)$ is the failure rate and $\mu(e)$ is the repair rate of edge e.[1] So, if edge e is *down* at time t, the probability that it will become *up* in the interval $[t, t + \delta t]$ is $\mu(e) \cdot \delta t + o(\delta t)$.

We remind the reader that the availability interpretation of the static probability has been discussed earlier, see Chapter 4, Section 3.

9.3.2 The mean stationary UP and $DOWN$ periods

Important parameters of the *dynamic* network are its *mean equilibrium UP* and *DOWN* periods, denoted as $\mu_{\mathbf{N}}(UP)$ and $\mu_{\mathbf{N}}(DOWN)$, respectively.

Let us define $\mu_{\mathbf{N}}(UP)$ more formally. ($\mu_{\mathbf{N}}(DOWN)$ is defined similarly.) Consider the sequence $\{U_i\}$, $i = 1, 2, ...$, of network *UP* periods. It can be proved that, as $i \to \infty$, the sequence $\{U_i\}$ converges in probability to the limit random variable U with mean value $\mu_{\mathbf{N}}(UP) = E[U] = \lim_{k \to \infty} E[U_k]$. (For details see [26, Chapter 3].)

Computing $\mu_{\mathbf{N}}(UP)$ and $\mu_{\mathbf{N}}(DOWN)$ is a difficult task. First, let us note, that in analogy with similar situation in renewal theory, the network stationary availability equals (see [26,20])

$$Av(\mathbf{N}) = R(\mathbf{N}) = \frac{\mu_{\mathbf{N}}(UP)}{\mu_{\mathbf{N}}(UP) + \mu_{\mathbf{N}}(DOWN)}. \qquad (9.3.1)$$

---

[1] note that $\lambda^\star(e)$ is component e failure rate and **not** the artificial birth rate $\lambda(e)$ in the evolution process

This formula alone, however, does not allow to calculate separately $\mu_{\mathbf{N}}(UP)$ and $\mu_{\mathbf{N}}(DOWN)$. Another important characteristic of the dynamic network in equilibrium is the so-called *stationary transition rate* $\Phi(\mathbf{N})$:

$$\Phi(\mathbf{N}) = \frac{1}{\mu_{\mathbf{N}}(UP) + \mu_{\mathbf{N}}(DOWN)}. \tag{9.3.2}$$

$\Phi(\mathbf{N})$ is a direct analogue of the equilibrium renewal rate for the alternating $UP - DOWN$ sequence of network states. In equilibrium, the average interval between two subsequent transitions $DOWN \rightarrow UP$ equals $\mu_{\mathbf{N}}(UP) + \mu_{\mathbf{N}}(DOWN)$. Therefore the $DOWN \rightarrow UP$ transition rate is exactly $\Phi(\mathbf{N})$. Obviously, knowing $R(\mathbf{N})$ and $\Phi(\mathbf{N})$ allows computing $\mu_{\mathbf{N}}(UP)$ and $\mu_{\mathbf{N}}(DOWN)$.

It turns out that $\Phi(\mathbf{N})$ can be expressed in the terms of the edge evolution process. First note that the transitions $DOWN \rightarrow UP$ take place only from so-called *border* states, see Chapter 5. Let us recall that the border state is such $DOWN$ state of the network which can be transformed into the UP state by a birth of a single edge. Denote by BD the set of all border states. For any border state S, denote by S^+ the set of all edges e such that $S+e \in UP$. Consider, for example the state $S = \{4, 5\}$ for the network on Fig. 9.2(a). This state is presented by super-state σ_{26} on Fig. 9.2(b). Clearly S is a border state and $S^+ = \{1, 2\}$.

Remark. To be more accurate, let us note that earlier we operated with "microscopic" definition of a state in a form of a binary vector. In these terms, the super-state σ_{21} represents, in fact, three "microscopic" border states $\mathbf{v}_1 = (1,1,0,0,0)$, $\mathbf{v}_2 = (1,0,1,0,0)$, and $\mathbf{v}_3 = (0,1,1,0,0)$. All these states have one property in common: adding edge 3 or 4 **to any one of them** transforms these states into the UP state of the network.#

So far we didn't know how to compute $\Phi(\mathbf{N})$. The key is provided by the following elegant formula (see [26, pp. 84, 110]):

$$\Phi(\mathbf{N}) = \sum_{S \in BD} P(S)\mu(S^+), \tag{9.3.3}$$

where $P(S)$ is the probability of the border state S, and

$$\mu(S^+) = \sum_{e \in S^+} \mu(e).$$

In the last formula, the sum is taken over the set S^+ of all edges e that are *down* and whose birth with rate $\mu(e)$ will lead from the border state S into the UP state.

The formula (9.3.3) can be explained heuristically as follows. Multiply the right-hand side by small time increment δt. Then the right-hand side is the probability that the network is (in equilibrium) in one of its border states $S \in BD$ *and* that a transition into UP takes place during this small interval. This is exactly what expresses the inverse of $(\mu_{\mathbf{N}}(UP) + \mu_{\mathbf{N}}(DOWN))$ multiplied by δt.

9.3.3 Estimation of $\Phi(\mathbf{N})$ for all-terminal connectivity

The number of border states even in a relatively small network is usually very large. Therefore we can only *estimate* the transition rate $\Phi(\mathbf{N})$. Let us explain here briefly the Monte Carlo scheme for estimating $\Phi(\mathbf{N})$.

Let us consider the all-terminal connectivity. The case of $s - t$ connectivity is more complicated technically, but not in principle. The difference between these criteria lies in the fact that for all-terminal connectivity, the border states are located on one (the $(r - 1)$-st) level of the turnip, where r is the highest level. All states on level $r - 1$ contain exactly two connected components. In the case of $s - t$ connectivity, the border states may be located on each level of the turnip and may consist of any number of components. For example, we can check on Fig. 9.2(b) that for the $s - t$ connectivity the state $S = \{5\}$ (which is associated with the super-state σ_{13}) is a border state and it represents three components. (A single node is also a component!)

Consider now the sum for $\Phi(\mathbf{N})$ in (9.3.3). Return to the Merging Process and regroup the terms in (9.3.3) in such a way that the "micro" border states are united into one super-state. Then we arrive at the following expression:

$$\Phi(\mathbf{N}) = \sum_{\sigma \in BD} P(\sigma)\mu(\sigma^+), \qquad (9.3.4)$$

where $\mu(\sigma^+) = \sum_{e \in \sigma^+} \mu(e)$ means the sum taken over the super-state σ of all edges e in *down* and whose birth rate $\mu(e)$ will lead from the border super-state σ to the UP state. As it was explained earlier, the static probability $P(\sigma)$ may be interpreted as the probability that the evolution process is in the super-state σ at the moment t_0. It should be emphasized that now we say "at the moment t_0" and not "before the moment t_0". (The fact that a random process **was** before the instant t_0 in a non-absorbing state does not mean, contrary to the absorbing state, that the process is at the instant t_0 in this state.) The UP state is an absorbing state while the border states (super-states) are not.

Denote by $\xi(\sigma)$ the random time for reaching the UP state. Now we have for $P(\sigma)$ the following formula:

$$P(\sigma) = P\{\xi(\sigma) \leq t_0\} - P\{\xi(\mathbf{N}) \leq t_0\}. \qquad (9.3.5)$$

Here the first term is the probability that the Merging Process is at t_0 in super-state σ or is in UP. The second term is the probability that the UP state was reached before t_0. The difference is therefore the desired probability that at t_0 the evolution process is in σ.

Now substituting (9.3.5) into (9.3.4) we obtain the following formula:

$$\Phi(\mathbf{N}) = \sum_{\sigma \in BD} [P\{\xi(\sigma) \leq t_0\} - P\{\xi(\mathbf{N}) \leq t_0\}] \cdot \mu(\sigma^+). \qquad (9.3.6)$$

To bring this formula to the form needed for Monte Carlo evaluation, let us introduce conditioning on the trajectories u leading from σ_0 to the UP state:

$$\Phi(\mathbf{N}) = \qquad (9.3.7)$$
$$\sum_{u \in U} p(u)[P\{\xi(\sigma) \leq t_0|u\} - P\{\xi(\mathbf{N}) \leq t_0|u\}] \cdot \mu(\sigma(u)^+).$$

We can write the formula in the above form because in the case of all-terminal connectivity, each trajectory u determines a unique border state $\sigma(u)$, the state from which it jumps into UP. (This does not take place for the T-terminal connectivity!). The sum in (9.3.7) has the form of mean value (compare with (9.2.8)) which is the key for Monte Carlo estimation on the turnip. For that purpose, we simulate trajectories u with the probabilities $p(u)$, as earlier, and compute for each simulated u the expression in the brackets (containing a difference of two convolutions) multiplied by $\mu(\sigma(u)^+)$.

9.3.4 Estimation of $\Phi(\mathbf{N})$ for T-terminal connectivity

Now let us turn to the calculation of the transition rate $\Phi(\mathbf{N})$, for a network with T-terminal connectivity. In general case, the border states can be located on any level of the turnip, and the transition into the UP state can take place not only from the super-state directly preceding the UP state, but also from several other super-states on the trajectory. For example, on Figure 9.3 the given trajectory contains the border states $\sigma_1, \sigma_2, \sigma_3, \sigma_4$. Now from (9.3.4), taking into account the above remark about border states location, we arrive at the following formula:

$$\Phi(\mathbf{N}) = \qquad\qquad\qquad (9.3.8)$$

$$\sum_{u\in U} p(u) \sum_{i=m_u}^{k_u} [P\{\xi(\sigma_i) \le t_0|u\} - P\{\xi(\sigma_{i+1}) \le t_0|u\}] \cdot \mu(\sigma_i(u)^+).$$

The inner sum in the above equation is taken over all border super-states belonging to the trajectory u. m_u is the first such state and k_u is the last. (So the super-state σ_{k_u+1} is in the UP.)

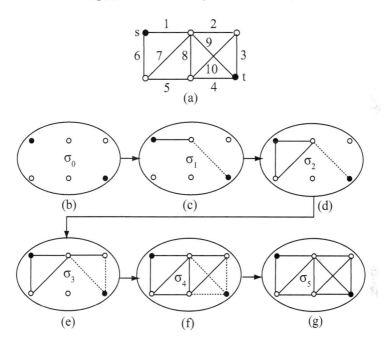

Figure 9.3: A fragment of a turnip and a trajectory leading into UP

Consider, for example, a 6-node, 10-edge network shown on Fig. 9.3(a). The nodes s, t are the terminals. Consider the trajectory $\sigma_0 \to \sigma_1 \to \sigma_2 \to \sigma_3 \to \sigma_4 \to UP$. It corresponds to the birth of edges, for example, in the following order: $1, 6, 2, 8, 9$. The super-states on the Fig. 9.3 are shown together with the corresponding closures. Note that the super-states $\sigma_1, \sigma_2, \sigma_3, \sigma_4$ all are border states. The dotted lines show the edges whose birth would immediately lead from the border state into the UP state. For example, σ_3 jumps into UP if any one of edges 9 or 3 makes the transition from *down* to *up*, with rates μ_9, μ_3, respectively. Let us consider the part of the sum in (9.3.8) corresponding to the above trajectory. We get:

$P(\sigma_1)\mu_9 + P(\sigma_2)\mu_9 + P(\sigma_3)(\mu_3 + \mu_9) + P(\sigma_4)(\mu_3 + \mu_4 + \mu_9)$.

Denoting by $\xi(\sigma_i)$ the random time of reaching the super-state σ_i we have:
$P(\sigma_1) = P(\xi(\sigma_1) \leq t_0) - P(\xi(\sigma_2) \leq t_0)$; $P(\sigma_2) = P(\xi(\sigma_2) \leq t_0) - P(\xi(\sigma_3) \leq t_0)$; $P(\sigma_3) = P(\xi(\sigma_3) \leq t_0) - P(\xi(\sigma_4) \leq t_0)$; $P(\sigma_4) = P(\xi(\sigma_4) \leq t_0) - P(\xi(\sigma_5) \leq t_0)$.

Taking into account that the first two terms in the sum have the same multiplier μ_9 we can "economize" on computing of appropriate convolutions:

$P(\sigma_1)\mu_9 + P(\sigma_2)\mu_9 = (P(\xi(\sigma_1) \leq t_0) - P(\xi(\sigma_3) \leq t_0))\mu_9$.

We give below the turnip algorithm for computing the transition rate $\Phi(\mathbf{N})$ for the case of T-terminal connectivity.

Algorithm 9.2 - TurnipFlow
1. **Put** $\Phi(\mathbf{N}) := 0$.
2. **Generate** trajectory $u = \sigma_0, ..., \sigma_k$ leading from the trivial super-state σ_0 to the super-state σ_k in UP.
3. **Find** the first $i = m_u$ so that σ_i is a border super-state, that is there exists $e \in \sigma_i^+$ leading into UP.
   (Note that all following super-states in the trajectory are also border super-states.)
4. **Calculate**
   $\Phi(\mathbf{N}) := \Phi(\mathbf{N}) + (P(\xi(\sigma_i) \leq t_0) - P(\xi(\sigma_{i+1}) \leq t_0))\mu(\sigma_i^+)$.
5. **Put** $i := i + 1$. **If** $i < k$ **Goto** 4.
6. **Repeat** 2-5 M times.
7. **Put** $\Phi(\mathbf{N}) := \Phi(\mathbf{N})/M$.

Note that the algorithm may be improved by using the "economizing" calculation mentioned in the above example.

9.3.5 Monte Carlo algorithm for the gradient

The notion of the reliability gradient vector $\nabla R = (\partial R/\partial p_1, ..., \partial R/\partial p_n)$ was introduced in Chapter 5. Let us remind that by formula (5.2.4) we have the following relationship between the gradient vector and the border state probabilities:

$$\nabla R \bullet \{q_1\lambda_1, ..., q_n\lambda_n\} = \sum_{\mathbf{v} \in BD} P(\mathbf{v})\Gamma(\mathbf{v}),$$

where

$$\Gamma(\mathbf{v}) = \sum_{[\mathbf{v} \in BD, \mathbf{v} + (0,...,0,1_i,0,...,0) \in UP]} \lambda_i.$$

Here $\Gamma(\mathbf{v})$ is the sum of all λ_i-s such that adding edge i to the fixed border state $\mathbf{v} \in BD$ "activates" this state and it becomes UP.

From the latter formula we can get the expressions for the components of gradient vector. In fact, what we will do now, is a generalization of the example 5.1.1. Let us *regroup* the terms of the above sum for $\Gamma(\mathbf{v})$ in such a way that for each element i we gather all border states which may be transferred into UP by adding this element. Then we arrive at the following expression:

$$\nabla R \bullet \{q_1\lambda_1, ..., q_n\lambda_n\} = \qquad (9.3.9)$$

$$\sum_{i=1}^{n} \lambda_i \sum_{[\mathbf{v}\in BD,\mathbf{v}+(0,...,0,1_i,0,...,0)\in UP]} P(\mathbf{v}).$$

From the latter we get the following formula for the gradient.

$$\nabla R = \{\frac{\partial R}{\partial p_1}, ..., \frac{\partial R}{\partial p_n}\}, \qquad (9.3.10)$$

$$\frac{\partial R}{\partial p_i} = \{q_i^{-1} \sum_{[\mathbf{v}\in BD,\mathbf{v}+(0,...,0,1_i,0,...,0)\in UP]} P(\mathbf{v})\}, i = 1, ..., n.$$

In this formula for $\frac{\partial R}{\partial p_i}$ we sum together the probabilities of all border states $\mathbf{v}$ which are "activated", i.e. move to UP by adding edge i.

Introducing conditioning on the trajectories of $u \in U$ and uniting the "microscopic" states into "macro" super-states, we arrive at the following formula for the partial derivatives:

$$\frac{\partial R}{\partial p_i} = \qquad (9.3.11)$$

$$q_i^{-1} \sum_{u\in U} p(u) \sum_{[j=m_u, \sigma_j+(0,...,0,1_i,0,...,0)\in UP]}^{j=k_u} P(\sigma_j), i = 1, ..., n.$$

Here U is the set of all trajectories u leading from the trivial super-state σ_0 into UP,
$u = (\sigma_0, ..., \sigma_{m_u}, ..., \sigma_{k_u}, \sigma_{k_u+1} \in UP)$,
$p(u)$ is the probability of trajectory u,
σ_{m_u} is the first border state on this trajectory, and σ_{k_u} is the last border state on it.

We give below the turnip algorithm for computing the gradient ∇R for the case of T-terminal connectivity.

Algorithm 9.3 - TurnipGradient

1. **Put** $\frac{\partial R}{\partial p_i} := 0, i = 1, ..., n$.
2. **Generate** trajectory $u = \sigma_0, ..., \sigma_k$ leading from the trivial
   state s_0 to the state s_k in UP.
3. **Find** the first $j = m_u$ so that σ_j is a border state,
   that is there exists $e \in \sigma_j^+$ leading into UP.
4. **Calculate** $Conv = P(\xi(\sigma_j) \le t_0) - P(\xi(\sigma_{j+1}) \le t_0))$;
   for each $e_i \in \sigma_j^+$ **calculate** $\frac{\partial R}{\partial p_i} := \frac{\partial R}{\partial p_i} + Conv$.
5. **Put** $j := j + 1$. If $j < k$ **Goto** 4.
6. **Repeat** 2-5 M times.
7. **For each** $i = 1, ..., n$ **put** $\frac{\partial R}{\partial p_i}/M$.

To make more clear the algorithm, let us take super-states σ_1 and σ_4 from the trajectory on Fig. 9.3. The super-state σ_1 contributes for the partial derivative $\frac{\partial R}{\partial p_9}$ the value $P(\xi(\sigma_1) \le t_0) - P(\xi(\sigma_2) \le t_0))$. The super-state σ_4 contributes for the partial derivatives $\frac{\partial R}{\partial p_9}, \frac{\partial R}{\partial p_3}, \frac{\partial R}{\partial p_4}$ the value $P(\xi(\sigma_4) \le t_0) - P(\xi(\sigma_5) \le t_0))$.

Remark: Border States. We have seen that the border states of the network appear in the context of solving two important problems: finding the network stationary UP and $DOWN$ periods (for networks with renewable components) and finding the network reliability gradient. The most natural way to design the corresponding Monte Carlo algorithms is using the edge evolution process where its trajectories allow automatic identification of the border states. The merging (closure) operation in Lomonosov's turnip increases the efficiency of the algorithms by grouping the "microstates" into the "macro" super-states. It is worth noting that network border states remain unchanged when the component reliability varies. In that sense the border states are network *invariants*, which makes all the corresponding algorithms suitable for arbitrary values of network component reliability.

In conclusion, we would like to note that using the turnip is not a unique possible way of finding the border states. In principle, any network *construction* process based on edge permutations, which starts from empty set and adds a component after component, see e.g. Section 6.3, leads to the UP state via a border state.#

9.4   Unreliable Nodes

Suppose that in our network the edges never fail and nodes, except the terminal nodes, are subject to failure. We are interested, as always, in the probability that the network is *UP*, i.e. all terminals are connected. (We exclude the trivial case when there are only two terminals and they are connected by an edge.) If we consider the *birth* process *on the nodes*, then it develops in the following way. First, we have an initial super-state σ_0 which contains only terminal nodes and no nodes which are subject to failure. We assume that there are no direct edges between terminal nodes.

Then nodes are born at random instants. As soon as two nodes a and b are born, and there is an edge $e = (a, b)$ in our network, these nodes get immediately connected by this edge. Similarly, if there is an edge connecting a terminal node to a newborn node, say a, this edge immediately becomes "alive". The node birth and evolution process goes on exactly as the above described process on edges, with only one distinction: there is no closure operation because there are no irrelevant nodes. As to the edges, they appear automatically as soon as their end-point nodes are born. In the process of appearance of new nodes, no irrelevant nodes exist.

To get used to the nodal birth and evolution process we advise the reader to solve the Problem 5 in the next section.

All above introduced notions - birth rates, trajectories, super-states, border states, the *UP* state, transition rates, sitting times - remain with no changes. All the theory and the simulation algorithms developed for edge evolution process (designed for unreliable edges) remain valid for node evolution process.

Consequently, all theoretical results and all algorithms can be translated in equivalent form from the "edge language" into the "node language".

9.5   Problems and Exercises

1. Below there is a turnip diagram for a bridge structure, with unreliable edges and $s - t$ connectivity as the *UP* state. One super-state, call it σ_{23}, is missing. Find it and complement the diagram.

2. Find the probability of trajectory $u = \{\sigma_0, \sigma_{11}, \sigma_{23}, \sigma_3\}$ and the CDF's of the sitting times in $\sigma_0, \sigma_{11}, \sigma_{23}$. Associate with edge i its birth time $\tau_i \sim Exp(\lambda_i)$. What is the mean transition time from σ_0 into *UP* along this trajectory?

3. Find, that given the trajectory u (Exercise 2), the probability P_0 that the time to reach UP is less or equal t. Find the probability P_1 to be in σ_{11} at time t.

4. Derive a formula for the probability that the evolution process on Fig. 9.4 will be absorbed by the super-state σ_{26}.

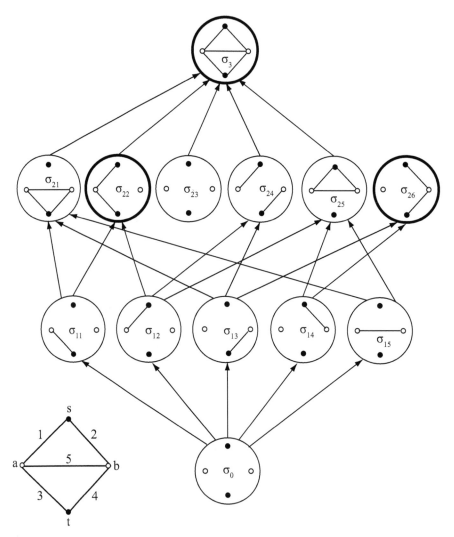

Figure 9.4: Turnip diagram for bridge structure. σ_{23} is missing. Bold circles belong to UP state

5. Evolution process on network nodes.

Consider a pentagon-type network with nodes numbered clockwise 1, 2, 3, 4, 5, and edges (1,2), (2,3), (3,4), (4,5), and (5,1). Nodes 1 and 3 are terminals. Edges do not fail, nodes are subject to failure (except the terminals). At $t = 0$ no node 2, 4, 5 exists. Each node i, $i = 2, 4, 5$ is supplied with its *birth time* $\tau_i \sim Exp(\lambda_i)$. If in the original network there is an edge $e = (i, j)$, then this edge has its "birth" at the instant of the appearance of both nodes i and j. The network is *UP* at the instant the nodes 1 and 3 become connected.

Describe the node evolution process, the transition rates, and the sitting times in the super-states. Sketch the turnip diagram.

Hint. The initial super-state is only nodes 1 and 3, with no edges. The first level of the turnip are three super-states with nodes $\sigma_1 = (1, 3, 5), \sigma_2 = (1, 3, 2), \sigma_3 = (1, 3, 4)$. σ_1 has edge (1,5), σ_2 has edges (1,2) and (2,3) and is already an *UP* state, σ_3 has edge (3,4). All second level super-states belong to *UP*.

6. Evolution process on a triangular network.

Consider a three node network. Nodes are denoted as a, b, c. The network has three edges $e_1 = (a, b), e_2 = (b, c)$, and $e_3 = (a, c)$. Edge failure probabilities are q_1, q_2, q_3, respectively. Nodes do not fail and the network is *UP* if at least two edges are *up*, i.e. all nodes communicate.

1. Construct the turnip diagram. Denote by σ_0 the "root" super-state with no edges and by σ_i the first level super-state with a single edge e_i. Find out all three trajectories u_1, u_2, u_3 leading from σ_0 to *UP*.

2. Determine the edge birth rates λ_i by the formula $\lambda_i = -log(q_i)$. ($t_0 = 1$). Find the distribution of the sitting time τ_0 in σ_0. Find the distribution of the sitting time τ_i in the super-state σ_i, $i = 1, 2, 3$.

3. Determine the probability $p(u_i)$ that the evolution process goes along u_i.

4. Compute the probability $P^* = P(\tau_0 \le 1) - P(\tau_0 + \tau_1 \le 1)$.
Hint. Use formula (5) from Appendix B.

5. Check that the static probability $p(\sigma_1)$ of the border state σ_1 satisfies the relationship

$$p(\sigma_1) = p(u_1) \cdot P^*.$$

Hint. $p(\sigma_1) = (1 - q_1)q_2 q_3$.

Chapter 10

Edge and Node Importance Calculation Using Spectrum

10.1   Introduction: Birnbaum Importance Measure

Let us remind the reader of the definition of Birnbaum Importance Measure (BIM) [3] of system component j, $j = 1, 2, ..., n$. If system reliability is a function $R = \Psi(p_1, p_2, ..., p_n)$ of component reliability p_j, we gave in Section 4.5 the following definition of BIM for component j:

$$BIM_j = \frac{\partial \Psi(p_1, ..., p_n)}{\partial p_j} = \Psi(p_1, ..., p_{j-1}, 1, p_{j+1}, ..., p_n) - \qquad (10.1.1)$$

$$\Psi(p_1, ..., p_{j-1}, 0, p_{j+1}, ..., p_n).$$

The BIM has a transparent physical meaning: it is the gain in system reliability received by replacing a down component j by an absolutely reliable one. Since it is also a partial derivative, it gives the value (approximately) of the system reliability increment δR resulted from component j reliability increment by δp_j. The first expression in the right-hand side of (10.1.1) is the reliability of a system in which component j is permanently replaced by an absolutely reliable one. The second term is the reliability of the system in which component j is permanently down. These expressions have an obvious interpretation for networks. If component j represents an edge $e_j = (a, b)$ then the first term is the reliability of a network in which the nodes a, b connected by e_j are "compressed" into one, while the second term is network reliability with edge e_j being permanently erased. If j represents

139

a node a, then the first term is system reliability if this note is absolutely reliable, and the second term represents system reliability where node a is eliminated, together with all edges incident to it.

The use of BIM in practice is very limited since usually we don't know the reliability function $\Psi(\cdot)$. It turns out however, that for the case of equal component reliability $p_j \equiv p$, there is a surprising connection between the network spectrum and the reliability function which allows to estimate the component BIM's without knowing the analytic form of the reliability function. This connection was first described in [22].

Remark. If we want to calculate the BIM analytically for the case of all $p_j = p$ we have to act as follows: **first**, calculate the partial derivative with respect to p_j and only **afterwards** plug in $p_j = p$.#

Our exposition is as follows. In Section 2 we define the so-called *cumulative spectrum* and derive formula for network reliability for the case of equally reliable edges. In Section 3 we derive our main result, the so-called BIM spectrum. In Section 4 we discuss how the component importance ranking depends on the value p. Section 5 contains a series of examples of edge and node importance.

10.2 Cumulative Spectrum

Example 10.2.1. Consider the network shown below:

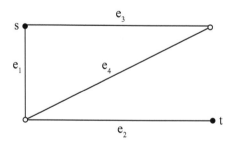

Figure 10.1: Network with unreliable edges. It is UP if s and t are connected

Consider an arbitrary permutation π of edge numbers, e.g. $\pi = (3, 1, 2, 4)$. Imagine the following network "construction" process. We start with a network without edges, and add one edge after another in the order mentioned in π, from left to right. So, first we add edge 3, then 1, then 2, and finally edge 4. Let us follow the state of the network in the process of its construction. Network is down after edge 3 is added; the same after edge 1 is added. Now if we add edge 2, the network becomes UP! Suppose now that we want to enumerate all UP states of this network. First, we may continue the construction process by adding edge 4, and assume that this edge is *up*; we can add the same edge and immediately erase it. So far we have two distinct UP states associated with π : (3+,1+, 2+, 4+) and (3+,1+,2+,4-) ("+" means *up*, "- " means *down*, respectively). Next we may find out how many permutations there are such that the associated construction process results in the UP state on the third step, <u>not earlier</u>. There are all in all 4!=24 permutations, and 14 out of them are of desired type. (Later we will say that all these 14 permutations have **anchor** equal 3.) From the first sight, we will have 14 different UP states, whose probabilistic weight will be p^4 and 14 UP states with weight p^3q each, $q = 1 - p$. Obviously, we already did a mistake since $14 \cdot p^4 > 1$ for $p > 0.52$. What is wrong? When we operate with a permutation, for us is important the *order* of edge appearance. But each of 24=4! permutations ending with four edges *up* defines in fact only a single event for the network whose probability is p^4. Therefore the contribution of one permutation with all edges being *up* is $14p^4/4!$ Similarly, every permutation with two edges *up* and two edges *down*, which produces network UP state, has $2! \cdot 2!$ copies differing by the order of appearance of the *up* and *down* components.

Let us finish the example. It can be proved that there are 4 permutations with anchor 2, and 6 permutations with anchor 4. This brings us to the following expression for the UP probability:

$$R = 4(\frac{p^2q^2}{2!2!} + \frac{p^3q}{3!1!} + \frac{p^4}{4!}) + \qquad (10.2.1)$$

$$14(\frac{p^3q}{3!1!} + \frac{p^4}{4!}) + 6\frac{p^4}{4!} = p^4 + 3p^3q + p^2q^2 = p^2 + p^3 - p^4.$$

The network on Fig. 10.1 is a series connection of edge 2 to edge 1 which is itself in parallel to edges 3 and 4. So, it is a series-parallel network and its reliability is easy to compute.#

Exercise. Verify the last formula for network reliability.

Definition 10.2.1.

Let π be a permutation of edges $e_1, ..., e_n$. Start with a network with all edges erased and add to it the edges in the order they appear in π, from left to right. Stop at the first edge when the network becomes UP. The ordinal number r of this edge is called the *anchor* of permutation π and denoted $r(\pi)$.#

For example, let $\pi = (e_5, e_3, e_2, e_4, e_1)$ for some network with 5 edges. Suppose that after adding e_5, e_3 the network is down, and after adding edge e_2 it becomes UP. Then the anchor of this permutation $r(\pi) = 3$.

Definition 10.2.2. Let x_i be the total number of permutations such that their anchor equals i. The set

$$C^\star = \{x_1, x_2, ..., x_n\} \tag{10.2.2}$$

is called the $C^\star$-*spectrum of the network.*#

For example, for the network on Fig. 10.1, the $C^\star$-spectrum is $\{0, 4, 14, 6\}$. The sum of all x_i equals to $n!$. Here $n! = 4! = 24$.

Remark. If we divide each x_i by $n!$ (normalize) we will get the $C^\star$-spectrum in a form of a discrete distribution, which was already introduced in Chapter 6.#

Definition 10.2.3.

Let $Y_b = \sum_{i=1}^{b} x_i$, $b = 1, 2, ..., n$.

Then the set $(Y_1, Y_2, ..., Y_n)$ is called the *cumulative* $C^\star$-*spectrum.*#

Theorem 10.2.1.

For all $p_i \equiv p$ network reliability can be expressed in the following form:

$$R = \sum_{i=1}^{n} Y_i \frac{p^i q^{n-i}}{i!(n-i)!}. \tag{10.2.3}$$

Proof.

Y_i is the total number of permutations which produce an UP state with i edges being up and $(n-i)$ being $down$. The probability of each of these states is $p^i q^{n-i}$. Among these states, there are copies arising from permutations of i up edges and $n - i$ $down$ edges among themselves. Each "true" UP state is counted, therefore, $i!(n - i)!$ times, which explains the appearance of $i!(n - i)!$ in the denominator. $\diamond$

10.3   BIM and the Cumulative $C^\star$-spectrum

Example 10.2.1 - revisited.

Let us consider all 14 permutations with the anchor equal 3. They are:

*(1,3,2,4)	(3,2,4,1)
*(1,4,2,3)	(3,4,2,1)
*(2,3,1,4)	*(4,1,2,3)

(2,3,4,1)	*(4,2,1,3)
*(2,4,1,3)	(4,2,3,1)
(2,4,3,1)	(4,3,2,1)
*(3,1,2,4)	
*(3,2,1,4)	

They all have the property that after three edges are added in the construction process, (and not earlier) the state becomes UP.

Let us consider the permutations with anchor equal 2. There are four of them:

$$*(1,2,3,4),*(1,2,4,3),*(2,1,3,4), *(2,1,4,3).$$

Now let us concentrate on one particular edge, say e_1, and count the number of permutations in which e_1 "took part" in the construction, i.e. it is on the first, second, or third position in all the above permutations (having anchor 2 or 3).

In the first group we have 8 such permutations marked by "*" and all 4 permutations in the second group, in total 12 permutations.#

Definition 10.3.1.

Denote by $Z_{i,j}$ the number of permutations satisfying the following two properties:

   (i) The network state constructed from the first i edges in the permutation is an UP state.

   (ii) Edge e_j is among the first i edges of the permutation. The collection of $Z_{i,j}$ values, $i = 1, 2, ..., n; j = 1, 2, ..., n$, is called Birnbaum Importance Measure Spectrum $(BIM \diamond S)$:

$$BIM \diamond S = \{Z_{i,j}, 1 \le i \le n; 1 \le j \le n\}. \tag{10.3.1}$$

Example 10.2.1 - continued. The table below presents the $BIM \diamond S$ for the network.

	$Z_{i,1}$	$Z_{i,2}$	$Z_{i,3}$	$Z_{i,4}$
$i = 1$	0	0	0	0
$i = 2$	4	4	0	0
$i = 3$	12	18	12	12
$i = 4$	24	24	24	24

Now we are ready to formulate the main result of this chapter.

Theorem 10.3.2.

The BIM for component e_j, $j = 1, 2, ..., n$, equals

$$BIM_j = \sum_{i=1}^{n} \frac{Z_{i,j} p^{i-1} q^{n-i} - (Y_i - Z_{i,j}) p^i q^{n-i-1}}{i!(n-i)!}. \qquad (10.3.2)$$

Proof.

Recall that

$$BIM_j = R(p_1, ..., p_{j-1}, 1_j, p_{j+1}, ..., n) - R(p_1, ..., p_{j-1}, 0_j, p_{j+1}, ..., n).$$

The number of permutations such that the first i components in it create an UP state *and* e_j is *up*, equals $Z_{i,j}$. Each fixed permutation counted in $Z_{i,j}$ creates an UP state having probability $p^{i-1} q^{n-i}$. Take into account that a specific state with i edges *up* and $(n-i)$ edges *down* is repeated $i!(n-i)!$ times in different permutations. Then the sum of probabilities for the states with i *up* and $(n-i)$ *down* components equals

$Z_{i,j} p^{i-1} q^{n-i} / (i!(n-i)!)$.

For the case of edge e_j being *down*, we obtain the expression for the corresponding probability as

$(Y_i - Z_{i,j}) p^i q^{n-i-1} / (i!(n-i)!)$, and the theorem follows.#

The main value of Theorem 10.3.2 lies in the fact that the importance measures can be estimated in the process of estimating the $C^\star$-spectrum. This is implemented in the following

Algorithm 10.1 - Computing BIM

1. **Initialize** all a_i and $b_{i,j}$ to be zero, $i = 1, ..., n; j = 1, ..., n$.
2. **Simulate** permutation $\pi \in \Pi_E$. (Π_E is the set of all edge permutations.)
3. **Find out** the minimal index of the edge $r = r(\pi)$ such that the first r edges in π create network UP state.
4. **Put** $a_r := a_r + 1$.
5. **Find** all j such that e_j occupies one of the first r positions in π, and for each such j put $b_{r,j} := b_{r,j} + 1$.
6. **Put** $r := r + 1$. **If** $r \le n$, **GOTO** 4.
7. **Repeat** 2 -6 M times.
8. **Estimate** $Y_i, Z_{i,j}$ via

$$\hat{Y}_i = \frac{a_i \cdot n!}{M}, \quad \hat{Z}_{i,j} = \frac{b_{i,j} \cdot n!}{M}. \qquad (10.3.3)$$

10.4 BIM and the Invariance Property

Suppose that for any pair of indices α and β, $\alpha \neq \beta$,

$$BIM_\alpha \leq BIM_\beta \quad \text{or} \quad BIM_\alpha \geq BIM_\beta \tag{10.4.1}$$

for **all** $p \in [0, 1]$.

If this is true we say that the BIM is invariant with respect to p. In practical terms it means that if e_α is more important than e_β for a particular value of p, then this remains true for all values of p. Therefore, the edge ordering according to their BIM's remains the same for all p values in the range [0,1].

Theorem 10.4.1.

Suppose we are given the $BIM \diamond S$ for our network. Let us fix two indices α and $\beta \neq \alpha$, and the corresponding $Z_{i,\alpha}$ and $Z_{i,\beta}$ values from the $BIM \diamond S$.

(i). If for all $i, i = 1, 2, ..., n$, $Z_{i,\alpha} \geq Z_{i,\beta}$, then $BIM_\alpha \geq BIM_\beta$ for all p values.

(ii). Suppose that (i) does not take place, and let k be the maximal index such that $Z_{k,\alpha} \neq Z_{k,\beta}$. Without loss of generality, assume that $Z_{k,\alpha} > Z_{k,\beta}$. Then there exists p_0 such that for $p \geq p_0$ $BIM_\alpha > BIM_\beta$.

Proof.

(i). Comparing the expressions (10.3.2) for BIM_α and BIM_β, we obtain after little algebra that

$$BIM_\alpha - BIM_\beta = \sum_{i=1}^{n} \frac{(Z_{i,\alpha} - Z_{i,\beta}) \cdot p^{i-1} \cdot q^{n-i-1}}{i!(n-i)!}, \tag{10.4.2}$$

which proves (i).

(ii). From the later expression we obtain that the same difference equals now to the sum of nonzero terms, when i runs from 1 to k. The largest of these terms has the factor $q = 1-p \to 0$ in the smallest degree, and therefore the sign of the whole expression will be determined by the sign of the largest term, for p close enough to 1. This proves (ii).#

Example 10.4.1. Parallel-series systems are BIM-invariant.

The system consists of n subsystems. The elements of subsystem $i, i = 1, ..., n$ have numbers $i_1, i_2, ..., i_{m_i}$ and respective reliability $p_{i1}, p_{i2}, ..., p_{im_i}$. All elements of one subsystem are connected in a series, and the subsystems are connected in parallel. The reliability of the i-th subsystem is

$$R_i = \prod_{j=1}^{m_i} p_{ij}, \tag{10.4.3}$$

and the reliability of the whole system is

$$R = 1 - \prod_{i=1}^{n}(1 - R_i). \tag{10.4.4}$$

Now let us calculate the partial derivative of R with respect to p_{ij}:

$$\frac{\partial R}{\partial p_{ij}} = \prod_{k \neq i}(1 - R_k) \prod_{k \neq j} p_{ik} = (1 - R)R_i(1 - R_i)^{-1}/p_{ij}. \tag{10.4.5}$$

Now let us set all $p_{ij} \equiv p$. The multiple of $(1 - R)/p$ becomes

$$f(m_i) = \frac{p^{m_i}}{1 - p^{m_i}}. \tag{10.4.6}$$

$f(m)$ is a *decreasing* function of m. So, the largest BIM will be attained for the subsystem with smallest number of elements. Suppose that $m_1 < m_2 < \ldots < m_n$. Thus the elements of the first subsystem are the most important, then follow the elements of the second subsystem, and so on. This ordering remains the same for all values of p and thus the parallel-series system follows the BIM invariance principle.

Example 10.4.2. Series-parallel systems are BIM-invariant.
    Suppose that our system consists of a series connection of k subsystems where the i-th subsystem consists of n_i components in parallel. Denote by S_i the index set of elements of the i-th subsystem. Elements are independent and element with index j has reliability p_j. System reliability has the following expression

$$R = \prod_{i=1}^{k}[1 - \prod_{j \in S_i}(1 - p_j)]. \tag{10.4.7}$$

    Suppose that element with index α is in the r-th subsystem, i.e $\alpha \in S_r$. Let us derive an expression for $\partial R/\partial p_\alpha$. It is seen from (10.4.7) that in the product the terms with $i \neq r$ remain unchanged, and the derivative of the r-th multiple becomes $\prod_{j \in S_r}(1 - p_j)/(1 - p_\alpha)$.
    Therefore,

$$\frac{\partial R}{\partial p_\alpha} = \prod_{i \neq r}[1 - \prod_{j \in S_i}(1 - p_j)] \cdot \frac{\prod_{j \in S_r}(1 - p_j)}{(1 - p_\alpha)}. \tag{10.4.8}$$

After setting $p_\alpha \equiv p$ and replacing $1 - p = q$ we will obtain

$$\frac{\partial R}{\partial p_\alpha} = q^{-1}\prod_{i=1}^{k}[1 - q^{n_i}]q^{n_r}. \tag{10.4.9}$$

It is already obvious that all elements of one subsystem have the same importance. Now suppose that $n_1 < n_2 < \ldots < n_k$. Then the largest q^{n_r} corresponds to the smallest n_1. Therefore, the elements of the subsystem with smaller number of components are more important in this case too. This conclusion remains true for any value of element reliability, which proves the BIM-invariance.#

10.5 Examples

Example 10.5.1. Four-dimensional hypercube network with four terminals.

In this example we consider a four-terminal hypercube H_4, see the figure below. It has 16 nodes and $2^4 \cdot 2 = 32$ edges. Nodes 1, 8, 9, and 16 are terminals. Hypercube configurations are often used in computer network design.

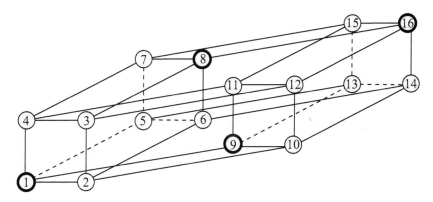

Figure 10.2: Nodes 1, 8, 9, and 16 are terminals

If an edge connects nodes α and β, it will be denoted as (α, β). Table 10.1 presents a fragment of simulation results based on simulating 10,000 permutations. Column 2 of the table gives the values of the *cumulative* $C^\star$-spectrum (denoted as a_i) for several i values. The next four columns present the corresponding values of $BIM \diamond S$. It is well seen from the table that the BIM's of edges (1,9) and (8,16) are equivalent and that their importance exceeds the importance of edge (1,2) which in turn is larger than the importance of edge (3,4). More detailed considerations (not presented here) allow to conclude that there are three groups of edges, in the descending order of their importance. The first group - edges(1,9) and (8,16), the second group

Table 10.1: Edge BIM's for H_4, 4 symmetric terminals

i	a_i	$b_{i,(1,9)}$	$b_{i,(8,16)}$	$b_{i,(1,2)}$	$b_{i,(3,4)}$
6	2	2	1	0	1
7	15	14	14	2	3
8	59	51	50	12	12
9	154	129	128	40	38
10	350	286	267	109	91
11	679	501	492	235	206
12	1333	886	904	525	438
13	2385	1478	1492	996	892
14	3723	2187	2210	1625	1547
15	5230	3012	3042	2502	2263

- all edges incident to the terminals 1, 8, 9, 16, and the third group - all the remaining edges.#

Table 10.2: Edge BIM's for H_4, three terminals

i	a_i	$b_{i,(1,9)}$	$b_{i,(9,10)}$	$b_{i,(8,16)}$	$b_{i,(3,4)}$	$b_{i,(7,8)}$
6	13	11	11	2	0	2
7	41	33	29	8	5	4
8	99	78	72	37	23	20
9	193	163	159	93	79	68
10	322	311	310	208	167	150
11	547	583	553	441	356	320
12	795	970	923	818	660	614
13	1195	1576	1527	1416	1173	1123
14	1351	2288	2234	2151	1845	1777
15	1435	3027	3033	2969	2681	2590

Example 10.5.2. Four-dimensional hypercube, 3 terminals, edge BIM's.

Consider now the same hypercube, with non symmetrically located terminals 1, 10, 16. Table 10.2 presents a fragment of the simulation results based on 10,000 replications. The notation is the same as in Table 10.1.

This case is less obvious because of non symmetrical positioning of the terminals. We can, based on the results, rank the edges in the following order:

$(1, 9) > (9, 10) > (8, 16) > (3, 4) > (7, 8)$.

(A simulation run with 100,000 replications leads to the same conclusion). An interesting feature of the data in Table 10.2 is that here we can not unite in one group, as it was in Example 1, all edges incident to one of the terminals. For example, edge (1,9) is ranked higher than edge (9,10). Note also that edges which are not incident to any terminal do not constitute a homogeneous group by their importance, as it was in Example 10.5.1. Such edges, for example, are (3,4) and (7,8).

Table 10.3: Node BIM's for H_4, three terminals

i	a_i	$b_{i,(2)}$	$b_{i,(9)}$	$b_{i,(15)}$	$b_{i,(7)}$	$b_{i,(6)}$
6	1161	800	805	208	140	208
7	2045	1712	1718	964	837	812
8	2483	2926	2926	2307	2088	1986
9	2052	4161	4165	3821	3569	3497
10	1097	5200	5204	5038	4889	4871
11	450	6091	6088	6028	5970	5964
12	158	6912	6904	6804	6909	6879
13	49	7696	7694	7699	7694	7692
14	0	8460	8460	8458	8465	8459
15	0	9232	9228	9227	9230	9228

Example 10.5.3. Node BIM's for H_4, 3 terminals.

Table 10.3 presents the results for 10,000 replications. As it is seen from this table, the nodes 2, 9, and 15 are ranked as $(2) = (9) > (15)$. All these nodes are the "neighbors" of terminals. Let us now check the last two columns of Table 10.3. We see that $b_{1,(7)} < b_{1,(6)}$, but for all $i > 1$ these inequalities are reversed. We have checked this fact by simulation with larger number of replications and obtained the same result. We have a combinatorial explanation of this fact: the number of short paths is greater for node 6, but the number of long paths is greater for node 7. Here the conditions of Theorem 10.4.1, (ii) hold (see Section 10.4), and therefore for

Table 10.4: Node BIM's for H_7, five terminals

i	$b_{i,(2)}/10^4$	$b_{i,(2)}/10^5$	$b_{i,(2)}/10^6$
20	0	0	0.00005
30	0.01500	0.01610	0.01854
40	0.135	0.13270	0.13559
50	0.29900	0.28900	0.29929
60	0.41300	0.41660	0.42126
70	0.51200	0.51400	0.51752
80	0.60100	0.60650	0.60722
90	0.69400	0.68610	0.69079
100	0.76500	0.77090	0.77191
110	0.84400	0.85160	0.85211
120	0.93800	0.93430	0.93405
125	0.97500	0.97580	0.97583

p greater than some p_0 node 7 becomes more important than node 6.

Example 10.5.4. Hypercube H_7, unreliable nodes. Importance estimates behavior for increasing number of replications.

Table 10.4 presents simulation results of importance measure for a fixed node in hypercube H_7 (128 nodes, 5 terminals, and 448 edges). Three simulation runs were made, for $K = 10^4, 10^5$ and $K = 10^6$ replications. It is seen from the table that the estimated BIM's display very small fluctuations with the increase of K.

10.6   Problems and Exercises

1. Solve the exercise in Section 2.

2. Investigate the connection between the $C^\star$ -spectrum and the D-spectrum introduced in Chapter 6.

3. The Fussel-Vesely importance measure (FVIM) for component i is defined in [18] as

$$FWIM_i = 1 - \frac{\Psi(p_1, ..., p_{i-1}, 0_i, p_{i+1}, ..., p_n)}{\Psi(p_1, ..., p_{i-1}, p_i, p_{i+1}, ..., p_n)}. \qquad (10.6.1)$$

The FVIM quantifies the decrement in system reliability caused by the failure of a particular component. Find the FVIM's for components of the network shown on Fig. 10.1.

4. The so-called FVIM-spectrum has been introduced in [22] as
$$FVIM \diamond S = V_{i,j} = \{Y_i - Z_{i,j}, 1 \le i \le n, 1 \le j \le n\}.$$
Prove that

$$FVIM_j = 1 - \sum_{i=1}^{n} \frac{V_{i,j} \cdot p^i \cdot q^{n-i-1}}{i!(n-i)!} \cdot R(p)^{-1}. \qquad (10.6.2)$$

Chapter 11

Optimal Network Synthesis

11.1 Introduction: Problem Formulation, Heuristic Algorithms

Optimal network synthesis is defined in a non formal way as the best possible design of the network (in terms of its reliability) under given constraints on the resources involved.

Example 11.1.1
Consider a four-terminal network shown on Fig. 11.1(a).

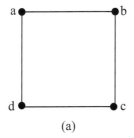
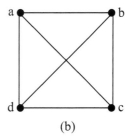

(a)   (b)

Figure 11.1: Four-terminal network with unreliables edges (left) and the network reinforced by two diagonal edges (right)

Nodes are reliable, edges fail independently and have equal failure probabilities $q = 0.01$. The network fails if the terminals become disconnected. Our "resource" is two "spare" edges each having failure probability $q_1 = 0.1$.

A spare edge can be put in parallel to any existing edge. The optimal synthesis problem in this toy example can be formulated as follows: find the best possible location for these two "spare parts" to minimize the network failure probability. For evaluating network failure probability we use the B-P approximation (see Section 4.4).

The original network has six minimal size min-cuts of size 2, and the main term of its failure probability equals $F \cong 6q^2 = 6 \cdot 10^{-4}$. If the spare edges are located on the diagonals, say as $e_1 = (a,c), e_2 = (b,d)$, then the network will have four min-cuts of size 3 and the corresponding failure probability will be $F_1 = 4 \cdot q^2 \cdot q_1 = 4 \cdot 10^{-5}$. It is easy to check that any other positioning of the spare edges would give larger failure probability. Thus the right network shown on Fig. 11.1(b) is the optimal solution to our problem.#

In this example the network is small and its failure probability can be easily approximated by the B-P formula or computed exactly. In fact, we act as if we have at our disposal an analytic expression for network reliability. In more realistic situations we don't have a formula or even a good approximation to the expression of system reliability. What we do have is the possibility to evaluate approximately, using simulation techniques, system reliability $R = \Psi(\mathbf{p})$ for a given vector $\mathbf{p} = (p_1, p_2, ..., p_n)$ of its component reliability.

Let us formulate a general heuristic procedure for "optimal" network synthesis.

We make the following assumptions:

1. The "initial" value of component reliability vector $\mathbf{p}^{(0)} = (p_1^{(0)}, p_2^{(0)}, ..., p_n^{(0)})$ is known.

2. For any $\mathbf{p}$, we have a method of estimating network reliability function $\Psi(\mathbf{p})$.

3. For any given $\mathbf{p}$, we have a method of estimating the reliability gradient vector $\nabla R = (\frac{\partial R}{\partial p_1}, ..., \frac{\partial R}{\partial p_n})$.

4. For any network component i, we know its "improvement cost function" $\psi_i(x, y)$, where $\psi_i(x, y)$ is the *cost* paid for changing component i reliability x to reliability $y, y > x$. Assume that the equation

$$1 = \psi_i(x, y) \tag{11.1.1}$$

can be solved for any $x \in [0, 1]$ with respect to y:

$$y = \psi_i^{-1}(x). \tag{11.1.2}$$

y is, therefore, the component i reliability obtained after investing one unit of "resource" if before this investment this component reliability was equal x.

5. There is a total resource C which can be spent for network optimal synthesis. C is an additive function of component improvement costs.

Our goal is to invest the resource C to obtain the maximal system reliability.

Now we formulate in a general form a heuristic algorithm for network synthesis under assumptions 1-5.

Algorithm 11.1 - Heuristic-1
1. Estimate $R^0 = \Psi(\mathbf{p}^0) = \Psi(p_1^0, p_2^0, ..., p_n^0)$.
   (Use any method available, e.g. the Turnip Algorithm, Chapter 9).
2. Estimate the partial derivatives $\nabla_i = \frac{\partial R}{\partial p_i}, i = 1, ...n$,
   evaluated at the point $\mathbf{p}^0$. (Use any method available.)
3. Compute the change d_i in component i reliability caused
   by investing into it one unit of resource:

$$d_i = \psi_i^{-1}(p_i^0) - p_i^0. \tag{11.1.3}$$

4. Find the component j for which the value

$$\nabla_i \cdot d_i \tag{11.1.4}$$

   is maximal.
5. "Invest" a unit resource into component j. **Recompute** the
   component reliability vector replacing p_j^0 by $p_j^1 = p_j^0 + d_j$
6. Repeat 2-5 until the whole resource C is exhausted.

Example 11.1.2.
Consider a series system of two renewable components with initial reliability $p_i = a_i/(a_i+b_i), i = 1, 2$, where a_i is the average *up* time and b_i is the average repair (*down*) time of component $i, i = 1, 2$.

Suppose that investing one unit of resource into component i reduces its repair time by factor $\alpha_i, \alpha_i < 1$. In other words, if initially the average repair time is b_i, then after investing one unit into it, the average time becomes $b_i \cdot \alpha_i$; after investing two resource units this time will be $b_i \cdot \alpha_i^2$. Suppose there are in total two resource units available.

The problem is to find the optimal resource allocation to provide maximum system availability. Since there are only three possibilities, we can find the best solution by enumeration: invest two units into component 1, or two units into component 2, or one unit into each of the components.

Let us carry out these simple calculations for $a_i = 1$, $b_i = 0.4, i = 1, 2$ and $\alpha_1 = 0.7, \alpha_2 = 0.5$.

It is easy to check that investing two units into component 1 will give the system availability $1/[(1 + 0.4 \cdot 0.49)(1 + 0.4)] = 1/1.674$. Investing two units into the second component results in availability $1/[(1 + 0.4)(1 + 0.4 \cdot 0.25)] = 1/1.54$. Finally, investing one resource unit into each component gives availability $1/[(1 + 0.4 \cdot 0.7)(1 + 0.4 \cdot 0.5)] = 1/1.536$.

So, the right answer is to invest one unit of resource into each component. Let us see what our algorithm Heuristic-1 will produce.

Since initially both components have equal availabilities, the partial derivatives are equal. The reliability increase $d_1 = 1/(1 + 0.4 \cdot 0.7) - 1/1.4 = 0.067, d_2 = 1/1.2 - 1/1.4 = 0.119$. Therefore the first resource unit should be invested into the second component. Now the component reliability vector is $\mathbf{p} = (1/1.4, 1/1.2)$ and therefore $\nabla_1 = 1/1.2, \nabla_2 = 1/1.4$.

Now compute $d_1 = \nabla_1 \cdot (1/1.28 - 1/1.4) = 0.056; d_2 = \nabla_2 \cdot (1/1.1 - 1/1.2) = 0.054$. Therefore, the second resource unit should be invested into the first component, which is the same result.#

Example 11.1.3. Optimal synthesis of a 3-node network.

We consider a network shown on Fig. 11.2.

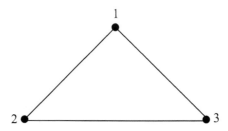

Figure 11.2: 3-node network with unreliables edges. All nodes are terminals

Its failure is defined as loss of connectivity. Edge reliability is x, y, z. Network is *UP* if at least two of its edges are *up*. This leads to the following expression of network reliability

$$R = xy + xz + yz - 2xyz.$$

It is easy to obtain the edge BIM's:

$$BIM_x = y + z - 2yz, \quad BIM_y = x + z - 2xz, \quad BIM_z = x + y - 2xy.$$

Let us assume that investing one resource unit into component having reliability p leads to the increase of its reliability by

$$d_p = 0.2 \cdot e^{-p}.$$

Thus, investing unit resource into a more reliable component gives less than investing it into a less reliable one. Suppose that d_x, d_y, d_z are "small". Then from the relationship

$$R(x + d_x, y + d_y, z + d_z) \approx R(x, y, z) + BIM_x \cdot d_x + \qquad (11.1.5)$$
$$BIM_y \cdot d_y + BIM_z \cdot d_z$$

it follows that the investment into a *single* component is most efficient if we choose the component which gives the maximal value among $Q_x = BIM_x \cdot d_x$, $Q_y = BIM_y \cdot d_y$, $Q_z = BIM_z \cdot d_z$. Suppose now that we have initially $x = 0.2, y = 0.3, z = 0.35$ and 10 units of resource. Let us synthesize our network, following the algorithm Heuristic-1.

Cycle 1. $R = 0.193$, the largest reliability increment is $Q_x = 0.072$, we improve the x component and set its reliability $x := 0.2 + 0.164 = 0.364$.

Cycle 2. $R = 0.265$, the largest reliability increment is $Q_y = 0.068$, we improve the y component and set its reliability $y := 0.3 + 0.148 = 0.448$.

Cycle 3. $R = 0.333$, the largest reliability increment is $Q_x = 0.067$, we improve the x component and set its reliability $x := 0.364 + 0.14 = 0.504$.

Let us skip the next 5 cycles and present

Cycle 9. $R = 0.765$, the largest reliability increment is $Q_y = 0.055$, we improve the y component and now its reliability $x = 0.889$.

Cycle 10. $R = 0.873$, the largest reliability increment is $Q_y = 0.052$, we improve the y component and now its reliability $y = 0.626$.

The final result is:

R=0.925, x=0.921, y=0.962, z =0.35.

Remark - nodes instead of edges: Suppose that the network on Fig. 11.2 is modified. Edge $(1, 2)$ is replaced by edges $(1, a)$ and $(a, 2)$, where a is a new node. Similarly, new nodes b and c are put between 2 and 3, and 1 and 3, respectively. Nodes 1,2,3 are declared terminals, new nodes a, b, c are subject to failure, but all edges are declared to be reliable. Node a, b, c reliability are x, y, z, respectively. The new network is *UP* if at least two nodes are *up*. The network reliability is exactly the same as in the previous case : $R = xy + xz + yz - 2xyz$. Therefore, we can repeat exactly the same cyclic procedure as above, but this time it will consider optimal **node synthesis**.

11.2   "Asymptotic" Synthesis

By asymptotic synthesis we mean optimal synthesis in situations where the network reliability function can be approximated by the asymptotic expression via the Burtin-Pittel formula (Section 4.4). The best way to demonstrate this method is to consider an example.

Example 11.2.1. Consider a two terminal network shown on Fig. 11.3.

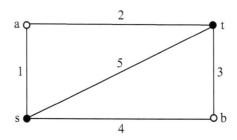

Figure 11.3: s and t are terminals. Network fails if $s - t$ connection is disrupted

s, t are terminals, reliability criterion is $s-t$ connectivity. Edges are subject to failures. Edge $e_i, i = 1, 2, ..., 5$ has failure probability $\alpha \cdot q_i$, and $\alpha \to 0$.

Following the derivation of the B-P approximation (see Section 4.4) it is easy to obtain that the main term in system failure probability F_0 is

$$F_0 = \alpha^3[q_1 q_4 q_5 + q_2 q_3 q_5 + q_1 q_3 q_5 + q_2 q_4 q_5], \qquad (11.2.1)$$

where each term corresponds to a minimal-size min cut (of size 3).

Assume further that, up to a multiple, component (edge) failure probabilities are $q_i = i, i = 1, 2, 3, 4, 5$. Then $F_0 = \alpha^3(20 + 30 + 15 + 40) = \alpha^3 \cdot 105$.

Now assume that we have at our disposal a certain resource by means of which we can convert any edge into absolutely reliable. In practice, it is enough to complement an edge by a similar one put in parallel. Then the reinforced edge e will have the failure probability $\alpha^2 \cdot q_e^2$, i.e. a quantity which is $o(\alpha \cdot q_e)$. For example, if we reinforce e_1, the main term F_0 will lose all terms containing q_1. In fact, it is formally equivalent to setting $q_1 = 0$. Obviously, reinforcing e_5 will be equivalent to reducing F_0 to zero, since now the main term of system failure probability will be of magnitude $O(\alpha^4)$. Our resource will be "spare" edges; spare for edge i costs c_i.

Let us denote by $F_0(i)$ the main term (omitting α^3) after reinforcing edge i. Obviously,

$$F_0(1) = 70, F_0(2) = 35, F_0(3) = 60, F_0(4) = 45, F_0(5) = 0. \qquad (11.2.2)$$

Thus the respective gain δF_i (i.e. the decrease in failure probability after reinforcing edge i, $i = 1, ...5$) will be

$$\delta F_1 = 35, \delta F_2 = 70, \delta F_3 = 45, \delta F_4 = 60, \delta F_5 = 105. \qquad (11.2.3)$$

Now assume that the costs are: $c_1 = 1, c_2 = 2.5, c_3 = 5, c_4 = 10, c_5 = 20$. In order to compare the effects of replacing different edges, let us compute the ratios $r_i = \delta F_i/c_i$. These ratios express the **gain in reliability per one cost unit**. In our example

$$r_1 = 35, r_2 = 28, r_3 = 9, r_4 = 6, r_5 = 5.25. \qquad (11.2.4)$$

Now let us formulate the following heuristic rule for choosing the best candidate for reinforcement:
reinforce the edge which has the **largest** r_i.

In our example, we start with reinforcing the edge e_1. Our gain will be 35 and our spending is $c_1 = 1$ of resource.

Now we have to rebuild the network. Node a now is placed at node s, and the network now has nodes s, t, c and edges $e_2 = (s, t), e_3 = (b, t), e_4 = (s, b)$, and $e_5 = (s, t)$, see Fig. 11.4.

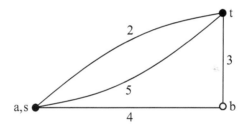

Figure 11.4: The network of Fig. 11.3 after edge e_1 becomes absolutely reliable

This new network has two min-cuts of size 3: one has edges (e_2, e_5, e_4), another - (e_2, e_5, e_3). Repeating the calculation similar to that already made for the original network, we obtain that further reinforcement of edge e_j reduces the failure probability by 70, 70, 30, 40, for $j = 2, 5, 3, 4$, respectively. The greatest gain in reliability **per unit cost** will be achieved by reinforcing edge e_2. After this, there will be a reliable $s - t$ connection and the system failure probability becomes $O(\alpha^4)$. The total cost of the two reinforcements is $c_1 + c_2$, which is optimal decision for a cost resource $C = 4.\#$

Now we formulate a heuristic algorithm for the "asymptotic synthesis" for given resource C.

Algorithm 11.2 - Heuristic-2

1. For a given network, **find** the B-P approximation to the failure probability F_0.
2. **Calculate** the reliability gain δF_i obtained from reinforcing edge i.
3. **Calculate** the quantities $r_i = \delta F_i / c_i$ and **find** the edge j with the largest value of r_i.
4. **Construct** a new network by replacing edge e_j by a reliable one. **Calculate** for it the B-P approximation $F_0^\star$ and set $F_0 := F_0^\star$.
5. **Redefine** the resource $C := C - c_j$
6. **IF** $C \leq 0$, or $F_0 = 0$, **STOP**; else **GOTO** 2.

11.3   Using Component Importance Measures

In this section we demonstrate how component importance measures can be used in designing a system with best reliability parameters. The following example illustrates our approach.

Example 11.3.1. Consider the network shown on Fig. 11.5. It has 7 nodes and 11 edges. We assume that edges are reliable and nodes 1,2,3,4,5 are subject to failures. Nodes s, t are terminals. Network fails if there is no connection between s and t.

Suppose that all nodes fail independently, and node probability to be *up* is p. Suppose that we are able to replace two of the existing nodes by another pair of more reliable nodes, each of which has reliability r, $r > p$. Our task is to choose the "best" pair of nodes for this replacement in order to maximize network reliability.

It is easy to establish, by means of a simple enumeration, that the network has the following 13 UP states:

$\{1,2\}, \{1,2,3\}, \{1,2,4\}, \{1,2,5\}, \{2,4,5\}, \{3,4,5\}, \{1,3,4\}, \{1,2,3,4\},$
$\{1,2,3,5\}, \{1,2,4,5\}, \{1,3,4,5\}, \{2,3,4,5\}, \{1,2,3,4,5\}.$
Now we can write the formula for network UP probability R:

$$R = p^5 + 5p^4 q + 6p^3 q^2 + p^2 q^3, \tag{11.3.1}$$

where $q = 1 - p$.
Suppose now that we replace nodes 1,2 for more reliable ones, with *up* probability r.

Then it is easy to get the following formula for network reliability R_{12}:

$$R_{12} = r^2 p^3 + 3r^2 p^2 q + 3r^2 pq^2 + r^2 q^3 + 2rp^3(1 - r) \tag{11.3.2}$$
$$+ 2rp^2 q(1 - r) + p^3(1 - r)^2.$$

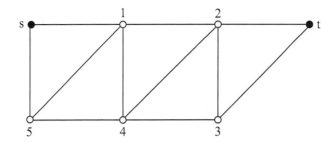

Figure 11.5: A network with reliable edges and unreliable nodes 1,2,3,4,5

Similarly, reinforcing nodes 3 and 4 (instead of 1,2) gives the reliability R_{34}:

$$R_{34} = r^2 p^3 + 3r^2 p^2 q + 2r^2 pq^2 + 2rp^3(1-r) + 3rp^2 q(1-r) \quad (11.3.3)$$
$$+ p^2 q(1-r)^2 + p^3(1-r)^2.$$

We leave as an exercise to prove that if $r > p$ then $R_{12} > R_{34}$. It means that reinforcing nodes 1,2 is preferable than reinforcing nodes 4,5.

On intuitive level, nodes 1,2 create the shortest path between s and t and its reinforcement is apparently better than the reinforcement of any other $s - t$ path. In general, comparing reliability functions for all possible pair of nodes is not a satisfactory method, even for small networks.#

Let us now formulate the problem considered in Example 11.3.1 in general form. Suppose that $k, 1 \le k \le n$ elements (nodes or edges) can be simultaneously replaced by more reliable ones. The problem is to choose the k candidates for the reinforcement in order to maximize the network reliability.

The solution of this problem is not trivial and involves the notion of *joint importance measure*. We will not investigate here this issue but instead provide a well-working heuristic method. This method uses the notion of BIM spectrum defined earlier in Chapter 10. For reader's convenience we remind some basic facts relevant to the BIM spectrum.

Definition 10.3.1

Denote by $Z_{i,j}$ the number of permutations satisfying the following two properties:
(i) The network state constructed from the first i edges in the permutation is an UP state;
(ii) Edge e_j is among the first i edges of the permutation.

The collection of $Z_{i,j}$ values, $i = 1, 2, ..., n$; $j = 1, 2, ..., n$, is called Birnbaum Importance Measure spectrum:
$BIM \diamond S = \{Z_{i,j}, 1 \le i \le n; 1 \le j \le n\}$.

Theorem 10.4.1.

Suppose we are given the $BIM \diamond S$ for our network. Let us fix two indices α and β, $\beta \ne \alpha$, and the corresponding $Z_{i,\alpha}$ and $Z_{i,\beta}$ values from the $BIM \diamond S$. Then

(i). If for all $i, i = 1, 2, ..., n$, $Z_{i,\alpha} \ge Z_{i,\beta}$, then $BIM_\alpha \ge BIM_\beta$ for all p values.

(ii). Suppose that (i) does not take place, and let k be the maximal index such that $Z_{k,\alpha} \ne Z_{k,\beta}$. Without loss of generality, assume that $Z_{k,\alpha} > Z_{k,\beta}$. Then there exists p_0 such that for $p \ge p_0$ $BIM_\alpha > BIM_\beta.\diamond$

Let us turn now to the problem of finding the best candidates for replacement and start with the case of replacing a single element. Suppose that the *up* probability of each element equals p and for the more reliable element the *up* probability is r. Let the two competing candidates for replacement be a and b. Let us introduce the following notations:

1. R_a and R_b - network reliability after replacing the elements a and b, accordingly.
2. S_i - the number of network UP states with i elements being *up* and $n - i$ elements being *down*.
3. X_i - the number of network UP states with i elements being *up* so that the element a is also *up*. Let $Y_i = S_i - X_i$. Let D_i and T_i $(D_i + T_i = S_i)$ - be the same quantities for the element b.

Suppose that $r > p$. Replacing the elements a and b we get, accordingly:
$R_a = \sum_{i=1}^{n} rp^{i-1}q^{n-i}X_i + \sum_{i=1}^{n} p^i q^{n-i-1}(1 - r)Y_i$,
$R_b = \sum_{i=1}^{n} rp^{i-1}q^{n-i}D_i + \sum_{i=1}^{n} p^i q^{n-i-1}(1 - r)T_i$.
Subtracting the second expression from the first, we obtain:
$R_a - R_b = \sum_{i=1}^{n} p^{i-1}q^{n-i-1}(X_i - D_i)(r - p)$.
From the latter expression we see that if for all $1 \le i \le n$ the inequality $X_i \ge D_i$ holds, then replacing element a is preferable. It is worth to mention that X_i (and also D_i) are related to the BIM spectrum value for the element a. Denote the corresponding value by $Z_{i,a}$. Then

$$X_i = \frac{Z_{i,a}}{i!(n - i)!}. \qquad (11.3.4)$$

If the inequality $X_i \geq D_i$ does not hold for all i, then let m be the maximal index such that $X_m \neq D_m$ and $X_m \geq D_m$. Then, similarly to the proof of the above mentioned theorem, it is possible to obtain that there exists p_0 such that for $p \geq p_0$, $R_a \geq R_b$.

The simultaneous replacement of k elements, $k > 1$, is more complicated and it is related to the *joint importance measure*. In order to simplify the exposition for replacing *several* network components, we suggest the following *heuristic* method.

Algorithm 11.3 - Heuristic-3
1. **Estimate** the BIM's values for all elements.
2. **Range** the elements by their BIM's values, from the "best" to the "worst".
3. **Take** the first "best" k elements and **replace** them by more reliable ones.

Example 11.3.1. Consider the hypercube H_3 network shown on Figure 11.6. It has 8 nodes and 12 edges. The nodes 1 and 6 are terminals. Edges are reliable, all nodes, except the terminals fail. Suppose now that we want replace two nodes by more reliable ones.

Compute the BIM spectrum by the algorithm from Section 10.3. It may be verified that the nodes are ranked in the following order:
(node 2=node 5)>(node 4=node 8)>(node 3= node 7).

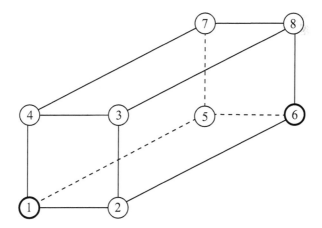

Figure 11.6: Hypercube H_3 network, nodes 1 and 6 are terminals

That is, the most important are the nodes 2 and 5, and so on. By the above recommended heuristic method, the best pair of nodes for reinforcement is $(2,5)$, which might seem obvious from Fig. 11.6. Computations which are not presented here confirm the reliability growth as the result of node replacement corresponds to the nodes ranking. For example, suppose that the node *up* probability equals 0.6. Then network reliability (computed by CMC) equals $R = 0.889$. If we replace the nodes 2 and 5 by more reliable, say with $p_2 = p_5 = 0.8$, then we get $R = 0.971$. Replacing nodes 2 and 4 gives reliability $R = 0.952$, replacing nodes 3 and 7 gives $R = 0.894.\#$

Remark. We can look at the previous example from a more general point of view. Suppose that in the hypercube network H_3 edges are reliable and nodes are subject to failure. We have to locate optimally *four* terminals at the hypercube nodes. By technical reasons, nodes 1 and 6 are already chosen as terminals. The problem is to locate optimally two remaining terminals. It obviously reduces to the previously considered situation with $r = 1$. We know the answer: the terminals must be placed at the nodes 2 and 5.$\#$

11.4   Problems and Exercises

1. Prove that $R_{12} > R_{34}$ for $r > p$, see (11.3.2), (11.3.3).

2. Prove, using the definition of the partial derivative of the reliability function $\partial R/\partial p_i$, (Section 10.1), that the maximal gain in reliability is attained by replacing component with $P(up) = p$ by a component with $P(up) = r = p + \delta$ corresponding to the maximal partial derivative.

3. Design a computer program to implement the Algorithm Heuristic-1 for the triangular network shown on Fig. 11.2.

4. The network has 5 nodes numbered 1,2,3,4,5 and edges (1,2), (2,3), (3,4), (4,5), and (5,1). Edges do not fail, nodes 1 and 3 are terminals. It is necessary to locate the third terminal in one of the nodes 2, 4, or 5. q is the failure probability of a non-terminal node.
a) What is the best location of the third terminal providing maximal reliability?
b) Solve a similar problem for a six node circular network in which two terminals are already located at the nodes 1 and 4.

Chapter 12

Dynamic Networks and Their Reliability Parameters

12.1   Introduction: Network Exit Time

We say that a network is dynamic if its elements (nodes or edges or both) "exist in time", i.e. change their state in time. More precisely, we associate with each element (unit) u a binary function $\xi_t(u)$, $t \geq 0$, with values 1 and 0 corresponding to the unit being *up* and *down*, respectively. Assume further that each $\xi_t(u)$ is a Markov process, i.e. $\xi_t(u)$ stays in the *up* and *down* states exponentially distributed random times with parameters $f(u)$ and $r(u)$, respectively. $f(u)$ is termed the failure rate and $r(u)$ - the repair rate.

The whole network is described by the set of random binary variables $V_t = \{u \in U : \xi_t(u) = 1\}$, representing the network state at moment t. Suppose that the network is UP at $t = 0$. The time τ of the first transition of the network into its $DOWN$ state is called the *exit time*:

$$\tau = \inf(s \geq 0 : V_s \in DOWN). \tag{12.1.1}$$

The main object of our study in this chapter is the distribution of the transition time $\phi_\tau(t)$. $\phi_\tau(t) = P(\tau \leq t)$ depends, of course, on the particular state $j \in UP$ of the network at time $t = 0$. We will try to find bounds from below and from above for $\phi_\tau(t)$.

Historically, the first step in this direction was made by considering the system at $t = 0$ in equilibrium UP state. Namely, suppose that

(i) at time $t = 0$ the network is in some of its UP states;

(ii) the system is already running for a very long time (formally, since $t = -\infty$);

(iii) nothing else is known about the behavior of the system before $t = 0$.

Then, assuming that the UP states are numbered as $1, 2, .., j, ..., n$, it follows from the ergodic renewal theory that the probability that the system is at $t = 0$ in particular state i equals

$$p_i^{erg} = \frac{p_i}{\sum_{j=1}^{n} p_j}, \qquad (12.1.2)$$

where p_i is the stationary probability of the state $i \in UP$. This probability (12.1.2) is called the *ergodic probability* of state i.

The following very important fact was established by Keilson [32].

$$1 - \phi_\tau(t) = P(\tau > t) \geq exp(-t/\mu_{\mathbf{N}}(UP)), \qquad (12.1.3)$$

where $\mu_{\mathbf{N}}(UP)$ is the *mean ergodic sojourn time* in UP. (A similar inequality is valid for the ergodic exit time from the $DOWN$ set, with obvious changes in the initial state distribution and with replacing $\mu_{\mathbf{N}}(UP)$ by $\mu_{\mathbf{N}}(DOWN)$.)

The inequality (12.1.3) has high theoretical value, but it was not used in practice because finding the mean *up* (or *down*) time for a stochastic system is by itself a very difficult problem.

An efficient approach to computing $\mu_{\mathbf{N}}(UP)$ and $\mu_{\mathbf{N}}(DOWN)$ (see Section 9.3), is based on using formulae (9.3.1) and (9.3.2):

(i) $Av(\mathbf{N}) = R(\mathbf{N}) = \mu_{\mathbf{N}}(UP)/(\mu_{\mathbf{N}}(UP) + \mu_{\mathbf{N}}(DOWN))$,

(ii) $\Phi(\mathbf{N}) = 1/(\mu_{\mathbf{N}}(UP) + \mu_{\mathbf{N}}(DOWN))$,

where $R(\mathbf{N})$ is the static network reliability and $\Phi(\mathbf{N})$ is the stationary transition rate. As it was described in Chapter 9, both these quantities can be efficiently calculated via Lomonosov's turnip.

Keilson's inequality is a simple and often quite accurate estimate of the exit time distribution function. Its main drawback is that it is only one-sided. The next section describes a general method of obtaining two-sided bounds on $\phi(t)$.

12.2   Bounds on the Network Exit Time

In this section we give a short description of an efficient method [46], of obtaining two-sided bounds for the exit time distribution function for networks with all-terminal connectivity criterion. Our method is based on using the

Lomonosov's turnip, a technique invented for calculating reliability parameters of *static networks*.

Before we describe our method in general form, let us consider an example which illustrates the main idea.

Example 12.2.1. Consider the network given on the Fig. 12.1(a). All nodes are terminals, the edges are subject to failures and subsequent renewals. Suppose that at $t = 0$, all edges are *down*. Suppose also that the following *true* process (Fig. 12.1(b)) of network transition from $DOWN$ to UP is developing as follows.

The changes in edge state appear at the instants $t_1 < t_2 < ... < t_7 < t_8$. At the moments $t_1, t_2, t_3, t_4, t_6, t_8$ the edges $4, 6, 2, 7, 1, 5$, respectively, are getting *up*, and at the moments t_5, t_7 the edges $4, 2$, respectively, are getting *down*. The network therefore becomes UP at the moment t_8.

Define now an *artificial* process (we call it *red process*) in which all events of edges going *down* are ignored. This process is shown on the Fig. 12.1(c).

Since the *down*-events are ignored, the first random moment when the *red process* goes up will be t_6. Denote it by σ_1 ($\sigma_1 = t_6$) and call it the *first guess* of the exit time τ.

To demonstrate the next guess, let us start a new *red process* from the true $DOWN$ state, at the moment $\sigma_1 = t_6$. We see that at this moment the edges 1,2,6,7 are *up* in the true process. Getting down of edge 2 at the moment t_7 is ignored. Finally, at the moment t_8 both the true and the *red* processes become UP. Clearly, the moment t_8 is the exit time. Denote $\sigma_2 = t_8$ and call it the *second guess* of the exit time τ.

We will use the above described random moments for constructing the upper and lower bounds.

Let us use the *red process* for constructing the first lower and first upper bound. As in the above example we will deal with the exit time from $DOWN$ to UP. Let σ_1 be the first *guess* of the exit time (we call it the *convergence time* of the *red* process starting at the equilibrium moment $t = 0$). Denote by $\varphi^{(1)}(t) = P(\sigma_1 \leq t)$ the distribution function of the convergence time of the red process. Note that from the convergence of the true process follows the convergence of the first *red* process. We have therefore the following inequality:

$$\varphi^{(1)}(t) \geq P(\tau \leq t) = \phi_\tau(t), \tag{12.2.1}$$

which is the first upper bound for the distribution function of the exit time.

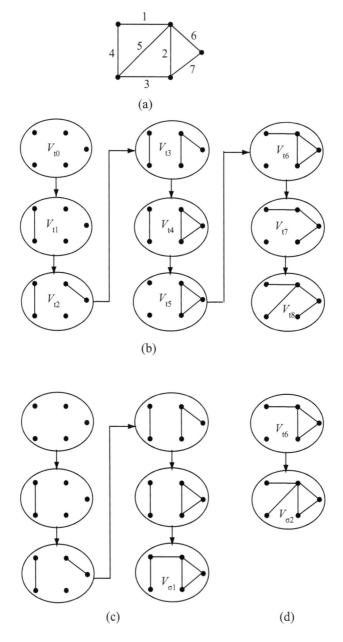

Figure 12.1: The network (a), the "true" process (b), the *red process*, first *guess* (c), the *red process*, second *guess* (d)

Table 12.1: Bounds on exit time for dodecahedron, $T=V$

q	$R(q)$	λt	$\Psi^{(1)}$	KB	$\phi^{(1)}$
0.5	0.024	1.0	0.08986	0.026307	0.85027
		0.5	0.08118	0.15201	0.44973
		0.25	0.06601	0.09045	0.17594
		0.1	0.04642	0.05135	0.06431
0.3	0.462	1.0	0.75499	0.93421	0.96856
		0.5	0.71222	0.81185	0.85498
		0.25	0.64269	0.68277	0.70647
		0.10	0.55499	0.56392	0.56952
0.15	0.919	1.0	0.99029	0.99619	0.99630
		0.5	0.97858	0.98242	0.98272
		0.25	0.96053	0.96225	0.96250
		0.10	0.94028	0.94068	0.94081
0.05	0.997	1.0	0.99986	0.99987	0.99987
		0.5	0.99940	0.99941	0.99941
		0.25	0.99872	0.99873	0.99872
		0.1	0.99798	0.99800	0.99800

Consider now the following event $A_t = \{(\tau \leq t) \cap (\sigma_1 = t)\}$, which means that the first convergence moment of the red process coincides with the exit time of the true process. Denote the probability of A_t by $\psi^{(1)}(t)$. Obviously, that

$$\psi^{(1)}(t) \leq \phi_\tau(t), \qquad (12.2.2)$$

since the exit time may coincide with any of the subsequent convergence moments of the red process. Thus we arrive at the lower bound for $\phi_\tau(t)$.

Let us turn now to the computational aspects of the upper bound $\varphi^{(1)}(t)$. Note that in the *red* process, unit u at $t = 0$ is *down* with probability $q(u) = \lambda(u)/(\lambda(u) + \mu(u))$. So, in the *red* process each unit u will be down at the moment t with probability $q(u)e^{-\mu t}$. From this we conclude that the distribution function $\varphi^{(1)}(t)$ coincides with the reliability of the *static* network, where each unit fails with the above defined probability $q(u)$: $\varphi^{(1)}(t) = R(\mathbf{N})$. Therefore, the upper bound may be computed by using the Lomonosov's turnip, see Chapter 9.

We omit here the computational aspects of the lower bound $\psi^{(1)}(t)$, because they are more complicated. Note that also $\psi^{(1)}(t)$ was computed by using the Lomonosov's turnip.

Remark. It has been proved in [46] that there is a way to iterate the *red* process and to obtain a monotone increasing sequence of lower bounds for $\phi_\tau(t)$ and a monotone decreasing sequence of upper bounds on $\phi_\tau(t)$. Moreover, it has been proved that the distance between the m-th lower bound and the m-th upper bound has magnitude of $O(t^{m+1})$ as $t \to 0$. Part of these bounds can be calculated using the "turnip technology". We omit here the details of the corresponding algorithms. The following numerical example demonstrates that for reliable networks, the first upper and the first lower bound on the exit time distribution are already very close to each other.#

Example 12.2.2: Dodecahedron (see Fig. 4.2), all nodes are terminals, edges are unreliable.

Table 12.1 presents simulation results of lower and upper bound for $\phi_\tau(t)$. We assume that all edges are identical, i.e. they have the same failure rate $\lambda(u) = \lambda$ and the same repair rate $\mu(u) = \mu$ (and therefore the same equilibrium probability of being *down*: $q = \lambda/(\lambda + \mu)$.) The network static reliability is denoted by $R(q)$. KB is the Keilson bound, $\varphi^{(1)}$ is the upper bound, and $\psi^{(1)}$ is the lower bound. All bounds were computed for different values of λt. The value t was fixed, $t = 1$ and the values of λ and q vary.

We may certainly conclude that the pair $[\psi^{(1)}, \varphi^{(1)}]$ provides a good estimate for the exit time distribution, even for not very reliable network. The total number of iterations was chosen to guarantee the relative error of the estimate within 1-2%. Taking 10,000 runs was enough for that purpose.

Chapter 13

Reliability of Communication Networks: Numerical Examples

13.1 Colbourn and Harms' Ladder-type Network

1. Ladder network, static case. Colbourn & Harms [7] investigated all-terminal reliability of a ladder-type computer-communication network shown on Figure 13.1

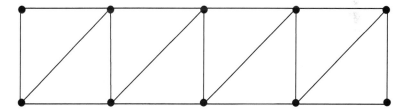

Figure 13.1: Colbourn - Harms network, all nodes are terminals

The unreliable elements are the edges. It was assumed that all edges are independent and have equal *up* probabilities $p(e)$ ranging from 0.1 to 0.99. Using the best up-to-date known lower (LB) and upper (UB) bounds, Colbourn & Harms compared these bounds with their exact analytic results $R_0(G,p)$. Table 13.1, columns 2 and 3 present their lower and upper bounds for various p values given in column 1. Our simulation results $\hat{R}_{MC}(G,p)$ obtained by the turnip algorithm (Section 9.2) are presented in column 5. They were obtained by averaging N independent replications, column 6. It

Table 13.1: Colbourn & Harms Bounds and Monte-Carlo Results

1	2	3	4	5	6
$p(e)$	LB	UB	$R_0(G,p)$	$R_{MC}(G,p)$	N
0.5	0.1052	0.2049	0.170	0.1578	100
0.6	0.2423	0.4415	0.3841	0.3648	100
0.7	0.45	0.7081	0.6409	0.6216	100
0.8	0.7081	0.8712	0.8548	0.8426	100
0.9	0.9302	0.9733	0.9710	0.9675	100
0.94	0.9791	0.9913	0.9908	0.9895	100
0.96	0.9922	0.9964	0.9962	0.9962	1000
0.98	0.99855	0.99914	0.99913	0.99912	1000
0.99	0.999714	0.999793	0.999791	0.999788	1000

is worth noting that in most cases, a relatively small number of replications N was needed to obtain an estimate $\hat{R}_{MC}(G,p)$ which lies within fairly tight bounds LB and UB.

2. Ladder network, renewable edges. Let us assume for the above considered ladder network that each edge e has failure rate $\lambda(e)$ and repair rate $\mu(e) = 1$, so that edge e has stationary *up* probability $p_{up}(e) = 1/(1 + \lambda(e))$. The parameter $\lambda(e)$ was determined for each value of $p(e)$ given in Colbourn & Harms [7] to satisfy the equality $p(e) = 1/(1 + \lambda(e))$.

When edges go *down* and *up*, similar process takes place for the whole network. Denote by $\mu_{\mathbf{N}}(UP)$ and $\mu_{\mathbf{N}}(DOWN)$ the average UP and $DOWN$ intervals for the network. We have already mentioned on several occasions that in order to find these intervals it is necessary to calculate the so-called stationary transition rate $\Phi(\mathbf{N})$. As it was shown in subsection 9.3.4, the estimation of Φ can be carried out by the **TurnipFlow** algorithm.

Table 13.2 presents the simulation results for various $p(e)$ values (column 1.) The estimates of the network reliability $\hat{R}_{MC}(G,p)$ are given in column 2, the transition rate estimates $\hat{\Phi}$ and the corresponding estimates $\hat{T}_{UP}$ and $\hat{T}_{DOWN}$ are in columns 5, 6, and 7, respectively. Column 4 gives the relative error δ in % of estimating Φ.

It is seen from Table 13.2, that the relative error lies within 1.3-8%, which is a relatively small value for rather moderate number of replications $N = 100 - 1000$ (column 3). As the network edges become more reliable ($p \to 1$),

Table 13.2: Ladder Network: Reliability, Transition Rate, Mean *UP* and *DOWN* Times

1	2	3	4	5	6	7
$p(e)$	$\hat{R}_{MC}(G,p)$	N	$\delta\%$	$\hat{\Phi}$	$\hat{\mu}_{\mathbf{N}}(UP)$	$\hat{\mu}_{\mathbf{N}}(DOWN)$
0.5	0.1578	100	2.5	0.808	0.195	1.042
0.6	0.3648	100	1.3	0.977	0.373	0.649
0.7	0.6216	100	0.6	0.760	0.818	0.498
0.9	0.9675	100	6	0.0732	13.22	0.444
0.94	0.9895	100	8	0.0228	43.33	0.46
0.96	0.9962	1000	3.3	0.00821	121.4	0.465
0.98	0.99912	1000	4	0.00184	543.3	0.481
0.99	0.999788	1000	4	0.000433	2308	0.490

$\mu_{\mathbf{N}}(DOWN)$ approaches 0.5. This can be explained by the fact that for R near 1, the dominant network failures are "cutting off" the upper-left or the lower-right nodes, and thus the corresponding repair rate is $\approx 2\mu(e) = 2$.

Table 13.3: Reliability of $s - t$ Connectivity of a Ladder Network

1	2	3	4	5
$\lambda(e)$	$\hat{Q}_{MC}$	$\hat{\Phi}$	$\hat{\mu}_{\mathbf{N}}(UP)$	$\hat{\mu}_{\mathbf{N}}(DOWN)$
10	0.02427	4.818	0.203	0.00504
20	0.05458	11.41	0.0829	0.00478
100	0.69608	106.9	0.00284	0.00651
200	0.69794	107.0	0.00283	0.00654
400	0.69937	107.4	0.00280	0.00651
600	0.69974	107.4	0.00280	0.00651
1000	0.70016	107.4	0.00279	0.00652

3. $s-t$ connectivity of a ladder network. Now we investigate the ladder network shown on Fig. 13.1 for $s - t$ connectivity assuming that only two nodes are terminals: s in the left upper corner and t in the right lower corner. It was assumed that $\mu(e) = 100$ for all edges, $\lambda(e) = 10$ for vertical and $\lambda(e) = 20$ for diagonal edges. The failure rate for horizontal edges changes from 10 to 1000, see column 1 in Table 13.3. The estimates of network failure probability $\hat{Q}_{MC}$ are given in column 2. Column 3 presents the estimate of transition rate $\hat{\Phi}$. All MC estimates were obtained for $N = 10,000$ replications. The corresponding relative errors for $\hat{Q}$ and $\hat{\Phi}$ are quite small and lie below 1%. Columns 4 and 5 present the MC estimates of $\mu_{\mathbf{N}}(UP)$ and $\mu_{\mathbf{N}}(DOWN)$. The estimation of Φ was made, similarly to the previous example, using the **TurnipFlow** algorithm.

As $\lambda(e)$ grows, the prevalent connection mode between s and t becomes a path without horizontal edges. Thus, with large probability, the $s - t$ connection is restored by renewing a single diagonal or vertical edge. This explains the fact that $\mu_{\mathbf{N}}(DOWN)$ lies below $[\mu(e)]^{-1} = 0.01$.

Table 13.4: $s - t$ Connectivity: 36 Nodes, 85 Edges, N=10,000

1	2	3	4	5
q_v	q_h	q_{diag}	$\hat{Q}_{s-t}$	$Q_{BP}(s-t)$
0.05	0.05	0.05	0.000271	0.00025
0.05	0.05	0.1	0.000483	0.00050
0.05	0.05	0.4	0.00189	0.002
0.05	0.05	1.0	0.00530	0.005

4. $s - t$ connectivity of a rectangular network.

Table 13.4 presents reliability simulation results of the $s - t$ connectivity for a 36-node, 85-edge rectangular network shown on Fig. 13.2.

q_v, q_h, q_{diag} are the edge failure probabilities for vertical, horizontal, and diagonal edges, respectively. The 4-th column presents network failure probability $\hat{Q}(s-t)$, and the fifth column - its Burtin-Pittel approximation, which turns out to be quite accurate. Note that this network has two minimal size min-cuts of three edges separating s and t from other nodes.

13.2   Integrated Communication Network (ICN)

13.2.1   General description

In this section we consider a more realistic example of an integrated communication network (ICN) designed for efficient coordination of several smaller networks.

A fast growing construction company has several construction sites spread over a large territory. These sites are represented by a network shown in the

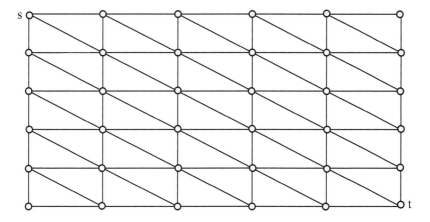

Figure 13.2: Rectangular network with 36 nodes and 85 edges. s, t are terminals, nodes do not fail

right upper corner of Fig. 13.3. There are nine such sites denoted as nodes of the network, numbered from 9 to 17. Node 13 represents the headquarter of the construction company. This company works in cooperation with three other organizations: the logistic material supply company LMS, the equipment repair and maintenance depot network RMD. and a computer network CN. LMS, RMD, and CN are themselves a network-type organization with branches located in different cities. Schematically, the LMS is represented by a local network shown on the left lower part of Fig. 13.3, RMD and CN are shown in the upper left part of Fig. 13.3 and right lower corner, respectively. The main offices of the LMS, CN, and RMD are presented by nodes 22 (of LMS), 1 and 8 (of RMD), and 27, and 34 of CN. Within each local network, the construction sites, local depot, computer centers, and maintenance and repair centers are connected by communication channels represented as network edges. In order to increase the overall efficiency of the construction works, the construction company decides to unite all companies into one Integrated Communication Network (ICN) by providing reliable connection between the headquarters (main offices) of all four specialized local companies, as shown on Fig. 13.3. As a whole, the ICN is now represented by a 34-node 54-edge network.

All communication channels (edges) are duplicated by telephone and cable connection, so that the edges are considered as being absolutely reliable. The network elements subject to failure are the nodes.

By the definition, the ICN is operating (i.e. is in the UP state) if there is provided the terminal connectivity between the headquarters and main offices $1, 8, 13, 22, 27, 34$.

In our terms, these nodes are declared to be terminals (which do not fail) and the network failure (the $DOWN$ state) is defined as the loss of the T-connectivity, $T = \{1, 8, 13, 22, 27, 34\}$.

13.2.2   ICN reliability

The D-spectrum. Our first task is estimating the D-spectrum of the ICN, see Chapter 8. We have simulated $N = 1,000,000$ random permutations of 28=34-6 nodes subject to failure and remembered for each permutation the minimal number of failed nodes causing the network failure. The simulation results are presented in Table 13.5. The minimal size min-cut has dimension 3 and $f_3 = 0.010969$.

Using formula (8.5.1) we find out that the number of minimal size min-cuts equals

$$W = f_3 \cdot 28!/3!25! = 35.9336.$$

The examination of all permutations leading to network failure on the third node failure reveals 36 such 3-node sets, which is in excellent agreement with the above value of W. Visual examination of the ICN scheme in order to find out all min-cuts of size 3 is not a trivial task.

Table 13.5: D-spectrum of the ICN

f_3	f_4	f_5	f_6	f_7	f_8
0.010969	0.035790	0.071126	0.108473	0.137634	0.149010

f_9	f_{10}	f_{11}	f_{12}	f_{13}	f_{14}
0.141963	0.120598	0.091331	0.061901	0.037152	0.019757

f_{15}	f_{16}	f_{17}	f_{18}	f_{19}	f_{20}
0.0009101	0.003608	0.001189	0.000329	0.000063	0.000006

Reliability and BP-approximation. Now, having the D-spectrum and using formula (8.4.4) we calculated the network failure probability $Q(G, p)$ as a function of node reliability p. The results are presented in Table 13.6. $\hat{R}(G, p) = 1 - \hat{Q}(G, p)$ (column 3) is the estimate of network reliability, and Q_{BP} (column 4) is the Burtin-Pittel approximation to the network failure probability computed by the formula

$$Q_{BP} = W \cdot (1 - p)^3. \tag{13.2.1}$$

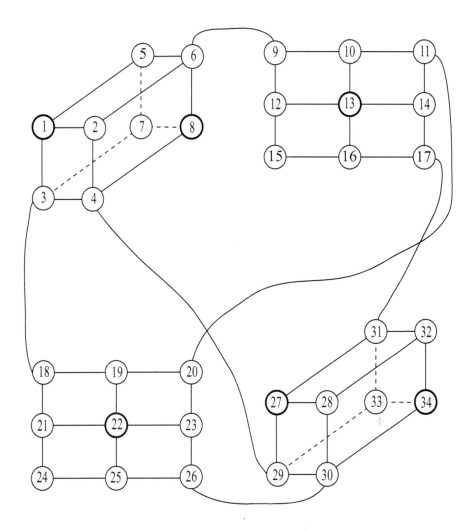

Figure 13.3: Scheme of the ICN: 34 nodes, 54 edges. Bold nodes are terminals

As it is seen from the Table 13.6, for $1 - p = q \leq 0.1$, this approximation is rather accurate, with relative error not exceeding 4%.

If the designers of the ICN want to guarantee that the network's UP probability $R \geq 0.9316$, the node reliability must be $p \geq 0.875$.

How accurate is our D-spectrum based estimates of network reliability?

Table 13.6: ICN Reliability

1	2	3	4
p	$\hat{Q}(G,p)$	$\hat{R}(G,p)$	Q_{BP}
0.50	0.943667	0.056333	——
0.60	0.80534	0.19466	——
0.70	0.54365	0.45635	——
0.8	0.233983	0.76617	——
0.85	0.11258	0.88742	
0.875	0.068433	0.931567	0.07031
0.9	**0.036338**	0.963662	**0.03600**
0.92	**0.018965**	0.981035	**0.01843**
0.93	**0.012781**	0.987219	**0.012335**
0.94	**0.008077**	0.991923	**0.007776**
0.95	**0.0046780**	0.995322	**0.004500**
0.99	**0.0000364**	0.999964	**0.0000360**

We used an upper bound for the variance of the reliability estimate which is based on formula (8.4.6). Delete the second sum in the expression for the variance and take $M = 10^6$. For $p = 0.95$ the standard deviation of the reliability estimate $\hat{\sigma}_R \leq 2 \cdot 10^{-5}$, for $p = 0.99$ $\hat{\sigma}_R \leq 3 \cdot 10^{-7}$.

Remark. We remind the reader that $Q(G,p)$ is computed using formula (8.5.5), where $\alpha = 1 - p$, or using an equivalent formula (8.4.5), with $F(t) = 1 - p$. The expression $Q = \sum_{x=1}^{m} C(x)(1-p)^x \cdot p^{(m-x)}$ is the network *static* failure probability for the case that all components fail independently and have equal failure probability $q = 1 - p$.

Obviously, system reliability $R(G,p) = 1 - Q(G,p)$ must be a *nondecreasing* function of component reliability p. We can see it from columns 1 and 3 of Table 13.6. Then obviously if $p \in [p_{min}, p_{max}]$, then system reliability $R(G,p) \in [R(G, p_{min}), R(G, p_{max})]$.

Moreover, these bounds are valid if component have *different* reliability within the interval $[p_{min}, p_{max}]$. For example, some components have $p = 0.9$, some - $p = 0.91$ and the rest - $p = 0.92$. Then system reliability will be in the interval $[0.963662, 0.981035]$, see Table 13.6.#

13.2.3 Network reinforcement

BP-approximation based heuristics. Our next task will be improving ICN reliability by reinforcing several nodes, in the spirit of the approach described in Chapter 11. Let us make the following assumptions.

1) We are able to reinforce at most 2 nodes;

2) A reinforced node becomes absolutely reliable.

Table 13.7 shows the node participation in the min-cuts (all cuts are of size 3). In total there are 36 cuts. Line 20, for example, tells that the cut number 20 consists of nodes $(28, 22, 17)$.

It follows from Table 13.7 that each of the nodes $3, 4, 6, 29, 31$ is involved in 9 cuts. Nodes $11, 17, 18, 20, 26, 30$ are involved in 8 cuts, Node 9 is a member of 7 cuts, and several other nodes are involved in a single cut.

The principal observation is the following: reinforcing a single node, say node 3, would reduce the number of min-cuts by 9, and thus, approximately, increase the reliability by $\delta R \approx 9(1 - p)^3$, see Section 11.2.

It is clear that the best single candidate for reinforcement is one of the nodes $3, 4, 6, 29, 31$. Suppose we reinforce node 3. Then we will be left with 36-9=27 cuts. Now eliminate all cuts containing node 3 from the list of 36 cuts. We have to repeat the whole procedure for the remaining cuts since there might be a considerable overlap of min-cuts, and eliminating all cuts with node 3 changes the contribution of the remaining nodes.

Let us spare rather elementary operations and present the final result. There are two most prospective nodes for reinforcement: node 4 belonging to 9 cuts and node 29 belonging also to 9 cuts after all cuts with node 4 are reinforced.

In total, our gain in system reliability will be

$$\delta R \approx 18 \cdot (1 - p)^3$$

Suppose we initially set $p = 0.9$ for each node and thus the ICN reliability is 0.963662 (see Table 13.6). Suppose we carry out the above described reinforcement of two nodes 4 and 29. They together reinforce 18 min-cuts. The gain in reliability will be $\delta R \approx 18 \cdot 0.1^3 = 0.018$, and the ICN reliability will be $R \approx 0.98166$. This is a rather considerable gain which otherwise might have been achieved by increasing all node reliability from $p = 0.9$ to $p \approx 0.922$. We should remember that all the above calculations are based on B-P approximation which are not 100% accurate and can produce an error. In our situation, the true value of ICN reliability obtained for

Table 13.7: Node participation in min-cuts

Cut No	Node No	Node No	Node No
1	(11)	(18)	(26)
2	(26)	(29)	(31)
3	(29)	(17)	(30)
4	(4)	(6)	(18)
5	(29)	(9)	(3)
6	(30)	(3)	(11)
7	(30)	(11)	(18)
8	(20)	(30)	(3)
9	(5)	(2)	(33)
10	(33)	(32)	(30)
11	(9)	(3)	(4)
12	(20)	(9)	(17)
13	(29)	(9)	(18)
14	(30)	(18)	(20)
15	(4)	(6)	(7)
16	(6)	(20)	(17)
17	(9)	(11)	(31)
18	(29)	(31)	(28)
19	(17)	(30)	(4)
20	(28)	(22)	(17)
21	(29)	(30)	(31)
22	(20)	(31)	(9)
23	(19)	(9)	(11)
24	(26)	(29)	(17)
25	(29)	(6)	(3)
26	(17)	(11)	(6)
27	(18)	(29)	(6)
28	(30)	(31)	(4)
29	(6)	(20)	(31)
30	(26)	(4)	(31)
31	(17)	(4)	(26)
32	(26)	(20)	(18)
33	(26)	(20)	(18)
34	(6)	(4)	(3)
35	(18)	(4)	(9)
36	(11)	(26)	(3)

all node reliability p=0.922 is $R^* = 0.982141$, which is rather close to our approximation $R = 0.981166$.

Table 13.8: ICN Node BIM Ranking

Node number	Node Rank
4,29	1
3,30	2
6,31	3
9,17,18,26	4
11,20	5
10,14,19,23	6
2,12,16,21,25,32	7
5,7,28,33	8
15,24	9

BIM-based heuristics. Let us now consider the node reinforcement policy based on their importance, in the spirit of Section 11.3. The cost factor will be ignored or, equivalently, assumed to be the same for all nodes. As it was suggested, for the case of replacing one element (node), the algorithm is as follows.

(1) Construct BIM's spectrum for all nodes.
(2) Range the elements by their spectrum, from the "best" to the "worst".
(3) Replace the most important appropriate element by more reliable.

The word "appropriate" in (3) reflects the situation in which some nodes can not be reinforced by some reasons. For example, in the ICN only nodes from LMS are allowed to be reinforced. In this case, after ranging the nodes, we choose for replacement the node with the highest BIM in the set $(18, 19, 20, 21, 23, 24, 25, 26)$.

The ICN node ranks are presented in Table 13.8. This ranking is not "absolute" in the sense that the same type of inequality ($>$ or $<$) for a pair of nodes not necessarily holds true for all i (for more details see Chapter 10).

Suppose that the *up* probability of all nodes is $p = 0.85$. The ICN reliability in this case equals $R \approx 0.887$. Table 13.9 presents the ICN reliability after reinforcing node k by a node with higher reliability p_r. For example, replacing node 4 by a node with $p_r = 0.9$ gives $R = 0.895$. We see from Table 13.9

Table 13.9: ICN Reliability After Reinforcing
a Node k

k	$p_r = 0.9$	$p_r = 0.95$	$p_r = 0.999$
4	0.895	0.902	0.910
29	0.895	0.902	0.910
9	0.894	0.900	0.906
18	0.894	0.900	0.907
10	0.889	0.891	0.892
14	0.890	0.891	0.893

Table 13.10: ICN Reliability After Pairwise Replacement

Node pair (k,m)	$p_k = p_m = 0.90$	$p_k = p_m = 0.95$	$p_k = p_m = 0.999$
(4,29)	0.903	0.920	0.934
(3,31)	0.903	0.917	0.932
(3,18)	0.902	0.916	0.933
(9,17)	0.900	0.912	0.923
(18,26)	0.900	0.912	0.923
(10,14)	0.891	0.893	0.896
(16,21)	0.890	0.891	0.894

that reinforcing node with higher rank is always preferable.

For a pair of nodes we use the following heuristic: reinforce the two nodes with maximal ranks. Suppose that all node *up* probability is 0.85.

Table 13.10 presents the ICN reliability after replacing a *pair* of nodes by more reliable ones.

In the first row we see the effect of replacing two most "important" nodes, 4 and 29. Table 13.10 demonstrates that for pairwise replacement our heuristic works quite well: the best results are obtained for a simultaneous replacement of the two nodes with highest BIM's.

Gradient-based heuristics.
Suppose now that the node *up* probabilities are not equal. For example, the probabilities are the following.
The RMD node (2,3,4,5,6,7) *up* probabilities are equal 0.6.

Table 13.11: Node Replacement by Gradient Algorithm

k	δp	$1^{st} : \frac{\partial R}{\partial p_k}$	$2^{nd} : \frac{\partial R}{\partial p_k}$	$3^{rd} : \frac{\partial R}{\partial p_k}$	$4^{th} : \frac{\partial R}{\partial p_k}$	$5^{th} : \frac{\partial R}{\partial p_k}$
3	0.35	**0.250**	—	—	—	—
4	0.35	0.242	**0.246**	—	—	—
6	0.35	0.191	0.208	0.152	**0.103**	—
9	0.25	0.157	0.147	0.128	0.082	**0.123**
17	0.25	0.209	0.253	**0.252**	—	—
18	0.15	0.151	0.239	0.194	0.181	0.169
26	0.15	0.204	0.196	0.142	0.138	0.133
29	0.05	0.169	0.135	0.215	0.199	0.156
30	0.05	0.185	0.179	0.144	0.117	0.132
31	0.05	0.171	0.214	0.185	0.293	0.195

The CC (9,10,11,12,14,15,16,17) *up* probabilities are equal 0.7.
The LMS node (18,19,20,21,23,24,25,26) *up* probabilities are equal 0.8.
The remaining nodes (28,29,30,31,32,33) CN have *up* probability equal 0.9.

Suppose that it is decided to replace **five** nodes for more reliable, with *up* probability 0.95. The question is which nodes are the best candidates for replacement to provide the maximal gain in reliability.

We suggest the following heuristic procedure based on component gradient, see Chapters 9,11.

(**1**). Compute component gradient for the initial network reliability
by the algorithm in 9.3.5.
(**2**). Choose the node having the maximal product of the partial derivative
times the node reliability increase, i.e. the maximal value of
$$\delta R = \delta p_k \cdot \frac{\partial R}{\partial p_k},$$
where $\delta p_k = 0.95 - p_k$ and p_k is the node k reliability.
(**3**). Set the reliability of the node found in (**2**) $p = 0.95$.
(**4**). Repeat (1-3) 5 times.

Table 13.11 presents the partial derivatives for some nodes (including the chosen nodes) on all computation stages. Column j, $j = 1, 2, 3, 4, 5$, presents the partial derivatives on the j-th step of the algorithm.

We see from the table that the nodes chosen for reinforcement were: 3, 4, 17, 6, 9. The reliability of the reinforced network is now 0.904, so that the gain in reliability equals $\delta R \approx 0.904 - 0.571 = 0.333$. To compare this result with other possibilities, let us take, for example, 5 nodes with highest BIM ranks (Table 13.8): 4, 29, 3, 30, 6. In this case the appropriate network reliability equals $R \approx 0.839$.

Remark. If for some node r $\partial R/\partial p_r = 0$, this means that node r does not affect the reliability of ICN terminal connectivity. Such nodes in our network are 15 and 24.

Appendix A: $O(\cdot)$ and $o(\cdot)$ symbols

When we investigate the behavior of function $f(x)$ as $x \to 0$ or $x \to \infty$, it is often desirable to compare *the order of magnitude* of $f(x)$ with the order of magnitude of some simple and/or known function $g(x)$. In this case we use the following notation.

1). If $|\frac{f(x)}{g(x)}|$ remains bounded as x tends to its limit, we write $f(x) = O(g(x))$, and say that the order of $f(x)$ is not greater in magnitude than the order of $g(x)$.

2). If $|\frac{f(x)}{g(x)}|$ tends to zero, we write $f(x) = o(g(x))$ and say that $f(x)$ has the order of magnitude smaller than $g(x)$.

3). If $\frac{f(x)}{g(x)}$ tends to one, then we write $f(x) \sim g(x)$ and say that $f(x)$ is asymptotically equal $g(x)$.

4). We also use the notation $O(1)$ to denote any function which does not exceed a certain constant, and the notation $o(1)$ to denote an expression (or function) which tends to zero.

All the above notations $O(f(x)), o(f(x)), O(1), o(1)$ have no meaning if we did not specify exactly the situation with the limiting behavior of x. We should tell exactly what happens with x, like x tends to zero, to infinity, etc.
Let us consider several examples.

1). $f(q) = q^2(1-q)^4, g(q) = q^2$ and $q \to 0$. Then $f(q) = O(q^2)$;
It is also correct to say that $f(q) = O(g(q))$, or $f(q) \sim g(q)$ as $q \to 0$.

2). Suppose $A(x) = a_0 + a_1 x + a_2 x^2$, $B(x) = b_0 + b_1 x$, all coefficients

185

a_i, b_i are not zeros.

Then $R(x) = \frac{A(x)}{B(x)} = O(1)$ as x tends to zero, and is $O(x)$ as x goes to $+\infty$.

3). Let $f(x) = 1 - \exp[-\lambda x]$, $x \to 0$, $\lambda > 0$ is fixed. Then $f(x) = O(\lambda x)$. If $g(x) = \exp[-\lambda x]$, then $g(x) = 1 - \lambda x + o(x)$ when $x \to 0$.

Appendix B: Convolution of exponentials

In this appendix we will consider the exact analytic expression for the convolution of r exponential distributions with parameters Λ_i, $i = 0, ..., r - 1$. It will be assumed that

$$\Lambda_0 > \Lambda_1 > \Lambda_2 > ... > \Lambda_{r-1}. \tag{1}$$

The fact that we take without proof is the following: If $\tau_i \sim Exp(\Lambda_i)$, $i = 0, 1, ..., r - 1$, then

$$P(\tau_0 + \tau_1 + ... + \tau_{r-1} \leq t) = 1 - \sum_{k=1}^{r} A_{r,k} e^{-\Lambda_{k-1}t}, \tag{2}$$

and

$$\sum_{k=1}^{r} A_{r,k} = 1. \tag{3}$$

The last formula follows from the fact the the convolution equals zero when we substitute $t = 0$ into (2).

The coefficients $A_{r,k}$ are found by the following recursive procedure.

$$A_{1,1} = 1; \tag{4}$$

$$A_{r+1,k} = A_{r,k} \cdot \frac{\Lambda_r}{\Lambda_r - \Lambda_{k-1}}, \quad k = 1, 2, ..., r;$$

$$A_{r+1,r+1} = 1 - \sum_{k=1}^{r} A_{r+1,k}.$$

Following the above recursive procedure, we would arrive at the following formula well-known in literature (see, e.g. Ross [44], p.299):

$$P(\sum_{i=0}^{r-1} \tau_i \le t) = 1 - \sum_{i=0}^{r-1} e^{-\Lambda_i t} \prod_{j \ne i} \frac{\Lambda_j}{\Lambda_j - \Lambda_i}. \tag{5}$$

Example. Let us consider a convolution of $r = 4$ exponentials with parameters $\Lambda_0 = 4, \Lambda_1 = 3, \Lambda_2 = 2, \Lambda_3 = 1$.

First,
$$P(\tau_0 \le t) = 1 - e^{\Lambda_0 t}.$$

Next
$$P(\tau_0 + \tau_1 \le t) = 1 - A_{2,1} e^{-4t} - A_{2,2} e^{-3t} = 1 - A_{2,1} e^{-\Lambda_0 t} - A_{2,2} e^{-\Lambda_1 t}.$$
By (4), $A_{2,1} = A_{1,1}\Lambda_1/(\Lambda_1 - \Lambda_0) = -3$. By (2), $A_{2,2} = 1 - A_{1,1} = 4$.
So, the convolution of the first two exponentials has the expression:

$$P(\tau_0 + \tau_1 \le t) = 1 + 3e^{-\Lambda_0 t} - 4e^{-\Lambda_1 t}. \tag{6}$$

Now $P(\tau_0 + \tau_1 + \tau_2 \le t) = 1 - A_{3,1} e^{-4t} - A_{3,2} e^{-3t} - A_{3,1} e^{-2t}$ and
$A_{3,1} = A_{2,1}\Lambda_2/(\Lambda_2 - \Lambda_0) = 3$ and
$A_{3,2} = A_{2,2}\Lambda_2/(\Lambda_2 - \Lambda_1) = -8$,
$A_{3,3} = 1 - A_{3,1} - A_{3,2} = 6$.

So, the convolution of the first three exponentials has the expression

$$P(\tau_0 + \tau_1 + \tau_2 \le t) = 1 - 3e^{-\Lambda_0 t} + 8e^{-\Lambda_1 t} - 6e^{-\Lambda_2 t}. \tag{7}$$

Finally,
$$P(\tau_0 + \tau_1 + \tau_2 + \tau_3 \le t) = 1 - A_{4,1} e^{-4t} - A_{4,2} e^{-3t} - A_{4,3} e^{-2t} - A_{4,4} e^{-t}.$$
$A_{4,1} = A_{3,1}\Lambda_3/(\Lambda_3 - \Lambda_0) = -1$,
$A_{4,2} = A_{3,2}\Lambda_3/(\Lambda_3 - \Lambda_1) = 4$,
$A_{4,3} = A_{3,3}\Lambda_3/(\Lambda_3 - \Lambda_2) = -6$, and
$A_{4,4} = 1 - A_{4,1} - A_{4,2} - A_{4,3} = 4$.
The convolution of four exponentials has the expression

$$P(\tau_0 + ... + \tau_3 \le t) = 1 + e^{-\Lambda_0 t} - 4e^{-\Lambda_1 t} + 6e^{-\Lambda_2 t} - 4e^{-\Lambda_3 t}. \tag{8}$$

Remark. We see from the above example that in the course of finding the convolution of the first r exponentials we obtain the convolution of the first $r - 1$ exponentials. This property is very convenient for calculations appearing in the turnip-flow algorithm since there we may need to find expressions of type $P(\sum_{i=0}^{r} \tau_i \le t) - P(\sum_{i=0}^{r+1} \tau_i \le t).\#$

Appendix C: Glossary of D-spectra

Table 1 presents edge D-spectra for a family of complete graphs with i nodes, denoted as $K_i, i = 5, 6, 7, 8, 9$, for all-node connectivity criterion. The data ware based on $10^5 - 10^7$ simulation runs. For example, if we want to find the 10-th element f_{10} of the spectrum for K_9, we look at the last column, row $r = 10$ and find $f_{10} = 0.0000108$

Table 2 presents edge D-spectra for a dodecahedron network (see Fig. 4.2). The first column is for the all node connectivity, the second for the terminal set $T = (1, 20)$, and the third for the terminal set $T = (1, 7, 8, 11, 16)$.

Table 3 presents the D-spectra for butterfly networks, for all node-connectivity. Notation $B_{wr}(24; 48)$ means a wrapped butterfly with 24 nodes and 48 edges, see Fig. 2.2(a). $B_{nw}(n, m)$ means a non-wrapped butterfly network with n nodes and m vertices.

Tables 4 and 5 present the D-spectra for hypercubes H_4, H_5, and H_6, obtained from simulating $10^6, 10^7$, and 10^8 permutations, respectively. For all hypercubes the criterion was all-terminal connectivity. H_4 has 16 nodes and 32 edges, H_5 has 32 nodes and 80 edges, and H_6 has 64 nodes and 192 edges.

The standard use of system spectrum is calculating system (static) reliability for the case that all components fail independently and have the same probability $q = 1 - p$ to be *down* and probability p to be *up*, see Chapters 6, 8, and Remark in Section 13.2.

Table 1: Edge D-spectra of complete graphs. ©IEEE 1991

r	K_5	K_6	K_7	K_8	K_9
4	0.023810				
5	0.095238	0.002050			
6	0.285714	0.010155	0.0001313		
7	0.595238	0.029820	0.0007801	0.0000067	
8		0.071937	0.0027032	0.0000456	0.0000003
9		0.155212	0.0072358	0.0001839	0.0000023
10		0.298834	0.0163476	0.0005710	0.0000108
11		0.431992	0.0329078	0.0014265	0.0000347
12			0.0615952	0.0031581	0.0000948
13			0.1087476	0.0062342	0.0002412
14			0.1822247	0.0117001	0.0005048
15			0.2777706	0.0205843	0.0010149
16			0.3095561	0.0344280	0.0019101
17				0.0555834	0.0033827
18				0.0867635	0.0058014
19				0.1305310	0.0094990
20				0.1867045	0.0151517
21				0.2406348	0.0231700
22				0.2214444	0.0 350933
23					0.0515021
24					0.0736922
25					0.1030831
26					0.1390625
27					0.1777469
28					0.2009639
29					0.1580374

Table 2: Simulated edge D-spectra of dodecahedron

r	$T = V$	$T = (1, 20)$	$T = (1, 7, 8, 11, 16)$
3	0.004916	0.000516	0.001234
4	0.015790	0.001733	0.004222
5	0.034064	0.004003	0.009603
6	0.060511	0.007903	0.018297
7	0.094845	0.014515	0.032249
8	0.136270	0.026398	0.054172
9	0.176819	0.045296	0.084642
10	0.201748	0.072431	0.125022
11	0.180148	0.105919	0.161231
12	0.094889	0.132139	0.170705
13		0.137760	0.139571
14		0.124864	0.093504
15		0.102451	0.054498
16		0.078137	0.028049
17		0.055645	0.013280
18		0.037199	0.005932
19		0.023832	0.002483
20		0.014196	0.000920
21		0.007905	0.000292
22		0.004212	0.000081
23		0.001916	0.000011
24		0.000749	0.000002
25		0.000239	
26		0.000042	

Table 3: Butterfly: simulated edge D-spectra; T=V

r	$B_{nw}(32;48)$ f_r	$B_{wr}(24;48)$ f_r	$B_{wr}(64;128)$ f_r	r	f_r
2	0.014151			32	0.025531
3	0.027989			33	0.027610
4	0.042251	0.000130	0.000006	34	0.029589
5	0.055542	0.000526	0.000020	35	0.031502
6	0.068253	0.001199	0.000071	36	0.033556
7	0.078894	0.002335	0.000129	37	0.035598
8	0.087785	0.004307	0.000212	38	0.037053
9	0.093999	0.007046	0.000322	39	0.038205
10	0.097654	0.010753	0.000504	40	0.039377
11	0.096963	0.015623	0.000733	41	0.040474
12	0.091938	0.021684	0.001050	42	0.040699
13	0.082169	0.029156	0.001291	43	0.040718
14	0.067577	0.034873	0.001794	44	0.040517
15	0.049631	0.048674	0.002299	45	0.039640
16	0.029592	0.059963	0.002775	46	0.038495
17	0.012612	0.072373	0.003413	47	0.037271
18	0.003000	0.084235	0.004129	48	0.035155
19		0.095270	0.005030	49	0.032486
20		0.103193	0.005901	50	0.029854
21		0.106647	0.007089	51	0.026795
22		0.101431	0.008245	52	0.023546
23		0.086905	0.009668	53	0.020337
24		0.063249	0.010786	54	0.016919
25		0.035391	0.012586	55	0.014006
26		0.011237	0.013860	56	0.010872
27			0.015708	57	0.008345
28			0.017718	58	0.006108
29			0.019578	59	0.004252
30			0.021441	60	0.002646
31			0.023370	61	0.001626
				62	0.000885
				63	0.000387
				64	0.000159
				65	0.000046
				66	0.000007

Table 4: H_4 and H_5 simulated edge D-spectra; $T = V$

	H_4; 10^6 trials		H_5; 10^7 trials	H_5; 10^7 trials	
r	f_r	r	f_r	r	f_r
4	0.000457			37	0.0570532
5	0.001774	5	0.0000013	38	0.0592518
6	0.004570	6	0.0000076	39	0.0602235
7	0.009166	7	0.0000176	40	0.0598683
8	0.016009	8	0.0000463	41	0.0580906
9	0.026141	9	0.0000975	42	0.0546688
10	0.040212	10	0.0001715	43	0.0497777
11	0.058031	11	0.0002823	44	0.0434966
12	0.080487	12	0.0004813	45	0.0360886
13	0.105428	13	0.0006681	46	0.0277087
14	0.131438	14	0.0009577	47	0.0192216
15	0.152050	15	0.0013438	48	0.0115109
16	0.160474	16	0.0018212	49	0.0053333
17	0.138589	17	0.0024286	50	0.0014290
18	0.075174	18	0.0032513		
19		19	0.0041148		
20		20	0.0052082		
21		21	0.0065489		
22		22	0.0081189		
23		23	0.0098713		
24		24	0.0119244		
25		25	0.0142721		
26		26	0.0168695		
27		27	0.0199113		
28		28	0.0231152		
29		29	0.0266203		
30		30	0.0304834		
31		31	0.0344016		
		32	0.0384775		
		33	0.0427933		
		34	0.0468323		
		35	0.0508296		
		36	0.0543587		

Table 5: H_6 simulated edge D-spectra; $T = V; 10^8$ trials

r	f_r	r	f_r	r	f_r	r	f_r
5	—-	37	0.00037336	69	0.00979215	101	0.02363795
6	10^{-9}	38	0.00043379	70	0.01048053	102	0.02268723
7	0.00000001	39	0.00050257	71	0.01121756	103	0.02158521
8	0.00000002	40	0.00057341	72	0.01193883	104	0.02043392
9	0.00000011	41	0.00065516	73	0.01272061	105	0.01914230
10	0.00000018	42	0.00074930	74	0.01347901	106	0.01775216
11	0.00000017	43	0.00084677	75	0.01429691	107	0.011632340
12	0.00000047	44	0.00096145	76	0.01511398	108	0.01487805
13	0.00000082	45	0.00108473	77	0.01595696	109	0.01340594
14	0.00000136	46	0.00121574	78	0.01680400	110	0.01193379
15	0.00000223	47	0.00136660	79	0.01768495	111	0.01048617
16	0.00000304	48	0.00152175	80	0.01854143	112	0.00905410
17	0.00000458	49	0.00171115	81	0.01937896	113	0.00771665
18	0.00000620	50	0.00189157	82	0.02020256	114	0.00647560
19	0.00000868	51	0.00210678	83	0.02105688	115	0.00533110
20	0.00001171	52	0.00233842	84	0.02186480	116	0.00429012
21	0.00001501	53	0.00257347	85	0.02260632	117	0.00338873
22	0.00002024	54	0.00284457	86	0.02337573	118	0.00260013
23	0.00002562	55	0.00313464	87	0.02401488	119	0.00194339
24	0.00003315	56	0.00343448	88	0.02464362	120	0.00140572
25	0.00004150	57	0.00376856	89	0.02520259	121	0.00097623
26	0.00005320	58	0.00412422	90	0.02568999	122	0.00065972
27	0.00006541	59	0.00449619	91	0.02606145	123	0.00041814
28	0.00008001	60	0.00490979	92	0.02638972	124	0.00024797
29	0.00009768	61	0.00534623	93	0.02656648	125	0.00013680
30	0.00011875	62	0.00578339	94	0.02662023	126	0.00006987
31	0.00014158	63	0.00629123	95	0.02662352	127	0.00002971
32	0.00016752	64	0.00680085	96	0.02645264	128	0.00001151
33	0.00018843	65	0.00734067	97	0.02613970	129	0.00000275
34	0.00023742	66	0.00791342	98	0.02573806	130	0.00000037
35	0.00027739	67	0.00850468	99	0.02516830		
36	0.00032487	68	0.00914229	100	0.02448057		

References

[1] Armstrong, M.J. 1995. Joint Reliability - Importance of Elements. *IEEE Trans. Reliab.*, **44**, (3), 408-412.

[2] Barlow, R. and F. Proschan. 1975. *Statistical Theory of Reliability and Life Testing: Probability Models*. Holt, Rinehart and Winston, Inc.

[3] Birnbaum, Z.W. 1969. On the Importance of Different Components in a Multicomponent System, *Multivariate Analysis-II*, ed. P.R. Krishnaiah, 581-592, New York: Academic Press.

[4] Bolshev, I.N. and N.V. Smirnov. 1965. *Tables of Mathematical Statistics*. Nauka (in Russian).

[5] Brassard, G. and P. Bratley. 1996. *Fundamentals of Algorithms*, Prentice Hall.

[6] Burtin, Y. and B.G. Pittel. 1972. Asymptotic Estimates of the Reliability of Complex Systems. *Engineering Cybernetics*, **10**(3), 445-451.

[7] Colbourn, C.J. and D.D. Harms. 1988. Bounding All-terminal Reliability in Computer Networks. *Networks*, **18**, 1-12.

[8] Cormen, Thomas H. 2001. *Introduction to Algorithms*, MIT Press.

[9] David, Herbert A. 1981. *Order Statistics*, 2nd ed., John Wiley and Sons, Inc.

[10] DeGroot, Morris H. 1986. *Probability and Statistics*, 2nd edition, Addison-Wesley Publishing Company.

[11] Elperin, T., Gertsbakh, I. and M. Lomonosov. 1991. Estimation of Network Reliability Using Graph Evolution Models. *IEEE Trans. Reliab.*, R-**40**, 572-581.

[12] Elperin, T., Gertsbakh, I. and M. Lomonosov. 1992. An Evolution Model for Monte Carlo Estimation of Equilibrium Network Renewal Parameters. *Probab. Engrng. Inf. Sci.*, **6**, 457-469.

[13] Everitt, B.S. 2002. *The Cambridge Dictionary of Statistics*, 2nd ed., Cambridge University Press.

[14] Fastor-Satorras, R. and A. Vespigniani. 2004. *Evolution and Structure of the Internet*, Cambridge University Press.

[15] *Fault Tree Analysis - Concepts and Techniques.* 1973 Proceedings of Nato Advanced Study Institute, Liverpool, England.

[16] Fishman, George S. 1996. *Monte Carlo. Concepts, Algorithms and Applications*, Springer.

[17] Ford, L.R. and D.R. Fulkerson. 1962. *Flows in Networks.* Princeton University Press.

[18] Fussel, J. and W.E. Vesely. 1975. How to Calculate System Reliability and Safety Characteristics, *IEEE Trans. on Reliability*, **24**(3), 169-174.

[19] Gertsbakh, I. 1989. *Statistical Reliability Theory*, Marcel Dekker, Inc.

[20] Gertsbakh, I. 2000. *Reliability Theory With Applications to Preventive Maintenance*, Springer.

[21] Gertsbakh, I. and Y. Shpungin. 1999. Product-Type Estimators of Convolutions, *Semi-Markov Models and Applications*, ed. J. Janssen and N. Limnios, Kluwer Academic Publishers, 201-206.

[22]. Gertsbakh, I. and Y. Shpungin. 2008. Network Reliability Importance Measures: Combinatorics and Monte Carlo Based Computations, *WSEAS Transactions on Computers*, Issue 4, **7**, 216-227.

[23] Gertsbakh, I. and Y. Shpungin. 2004. Combinatorial Approaches to Monte Carlo Estimation of Network Lifetime Distribution. *Appl. Stochastic Models Bus. Ind.*, 20, 49-57.

[24] Gertsbakh, I. and Y. Shpungin. 2003. Combinatorial and Probabilistic Properties of M.V. Lomonosov's "Turnip". *Proc. Internat. Conference on Reliability in Communication*, Riga, 14-21.

[25] Gertsbakh, I. and H. Stern. 1978. Minimal Resources for Fixed and Variable Job Schedules, *Operations Research*, **26** (1), 68-85.

[26] Gnedenko, B.V., Y. Belyaev and A.D. Solovyev. 1983. *Mathematical Reliability Theory.* Moscow, "Radio i Sviaz"' (in Russian).

[27] Gnedenko, Boris and I. Ushakov. 1995. *Probabilistic Reliability Engineering.* John Wiley and Sons, Inc.

[28] *Handbook of Reliability Engineering.* 1993. ed. Igor Ushakov, John Wiley and Sons, Inc.

[29] Hui, King Ping. 2007. Monte Carlo Network Reliability Ranking Estimation, *IEEE Trans. Reliability,* **56**,(1), 50-57.

[30] Hui, K-P, Bean, N., Kraetzl, M. and Dirk P. Kroese. 2005. The Cross-Entropy Method for Network Reliability Estimation. *Annals of Operations Research,* **134**, 101-118.

[31] Hui, K-P. et al. 2003. The Tree Cut and Merge Algorithm for Estimation of Network Reliability, *Prob. in Engineering and Information Sciences,* **17**, 23-45.

[32] Keilson, J. 1979. *Markov Chain Models - Rarity and Exponentiality,* (Applied Math. Sci, **28**). Springer, New York.

[33] Kordonsky , Kh. B. and I. Gertsbakh. 1976. Using Entropy Criterion for Job-Shop Scheduling Algorithm. *Naval Res. Logistics Quarterly,* 185-192.

[34] Kruskal, J.B. 1956. On the Shortest Spanning Tree and the Traveling Salesman Problem, *Proceedings of the AMS,* **7**, 48-50.

[35] Lisniansky, A. and G. Levitin. 2003. *Multi-State System Reliability,* World Scientific Publishing Co., Ltd.

[36] Lomonosov, M. 1974. Bernoulli Scheme With Closure, *Problems of Information Transmission (USSR),* **10**, 73-81.

[37] Lomonosov, M. 1994. On Monte Carlo Estimates in Network Reliability, *Probab. Engnrg. Inf. Sci.,* **8**, 245-264.

[38] Lomonosov, M. and Y. Shpungin. 1999. Combinatorics and Reliability Monte Carlo, *Random Structures and Algorithms,* 329-343.

[39] Mitzenmacher, M. and E. Upfal. 2005. *Probability and Computing,* Cambridge University Press.

[40] Moore, E.F. and C.E. Shannon. 1956. Reliable Circuits Using Less Reliable Relay. *J. of Franklin Institute*, **262**, 1, 191-208; **262**, 2, 281-297.

[41] *Network Reliability: Experiments with a Symbolic Algebra Environment (Discrete Mathematics and Its Applications)*. 1995. editors D. Harms, M. Kraetzl, Charles J. Colbourn. CRC Press.

[42] Provan, J. and M. Ball. 1982. The Complexity of Counting Cuts and of Computing the Probability that a Graph is Connected. *SIAM Journal of Computing*, **12**, 777-787.

[43] Ross, Sheldon M. 1997. *Simulation*, 2nd ed., Academic Press.

[44] Ross, Sheldon, M. 2007. *Introduction to Probability Models*, 9th ed, Academic Press, Inc.

[45] Rubinstein, R.Y. and D.P. Kroese. 2007. *Simulation and Monte Carlo Method*, 2nd ed., John Wiley and Sons.

[46] Shpungin, Y. 1997. Combinatorial and Computational Aspects of Monte Carlo Estimation of Network Reliability, Ph.D. Dissertation, Dept. of Math. and Comp. Sci., Ben-Gurion University, Beersheva.

Index